A. Cristina Correia Manicardi
Sueli Rizzutti
Monica C. de Miranda

Development of a battery of executive functions

A. Cristina Correia Manicardi
Sueli Rizzutti
Monica C. de Miranda

Development of a battery of executive functions

For school children

ScienciaScripts

Imprint

Cover image: www.ingimage.com

This book is a translation from the original published under ISBN 978-613-9-73407-8.

Publisher:
Sciencia Scripts
is a trademark of
Dodo Books Indian Ocean Ltd. and OmniScriptum S.R.L publishing group

120 High Road, East Finchley, London, N2 9ED, United Kingdom
Str. Armeneasca 28/1, office 1, Chisinau MD-2012, Republic of Moldova, Europe
Printed at: see last page
ISBN: 978-620-6-21166-2

I dedicate this work to my parents, for being people equally beautiful and admirable in essence, for the stimuli that pushed me to seek new life every day and for having granted me the opportunity to fulfil myself even more.

THANKS

I begin by thanking God, who was always by my side during this journey. Many times the path became tortuous and I thought of giving up, but He gave me two characteristics that are embedded in my soul: persistence and determination! However, I would not have got this far without the help of some angels that He sent me.

I thank my parents, Regina Helena and Admilson, who have always been with me, teaching me, supporting me, loving me unconditionally and believing in my potential. I love you!

A special thanks to my great partner Cleusa Manicardi, for the patience (it took a lot), dedication, for the care and all the love and affection, sparing no effort for me to reach this stage of my life.

Thank you to all my family and friends, in particular my sisters, my nephews, my great-nephews and my beautiful goddaughters, who understood (sometimes not so much) my absence on many commemorative dates.

I would like to thank my supervisor, Professor Sueli Rizzutti, for believing in me and accepting me as a student.

To Professor Doctor Monica Carolina Miranda, who was more, much more than a co-supervisor, for her competence, for inviting me to participate in this brilliant work, for encouraging me, supporting me whenever I needed, for the affection and care with my health that, sometimes, I neglected.

Still in the academic field, I must thank Professor Doctor Sabine Pompéia, an example that I will always take with me, as a person and a professional, thank you for believing in my work.

I thank Raquel Antonio Luna, for the partnership and for the pertinent suggestions throughout the construction of this work.

There is no way to describe the immense encouragement I have had from my colleagues Fátima Ferreira Dias, Carolina Nikaedo and Luzia Flavia Scaramuzza, since the beginning of this journey.

I thank everyone who is part of the Núcleo de Atendimento Neuropsicológico e Interdisciplinar (Nani) family: thank you all, you are very special. Thank you for the affection and the words that I have heard over all these years!

To professors Elizeu Coutinho de Macedo, Mauro Muszkat, Alessandra Gotuzo Seabra and Lucia Maria Gonzales Barbosa, who accepted to be part of my qualification and defense boards, for their suggestions and significant analyses to which I tried to attend in the definitive version of the work.

I could not forget Nani's secretary, Ariane, for her patience and immense help at various times.

Finally, I thank the Postgraduate Programme in Education and Health in Childhood and Adolescence and all the teachers of the department who fight for a worthy education and quality teaching. Thank you very much!

"And I've learned that if you always depend
Of so many, many, different people
Every person is always the imprints of the daily lessons of so many other people.
It is so beautiful when we understand that we
are so many people wherever we go.
It's so beautiful when we feel
Who is never alone
No matter how much I think I am...". (Paths of the Heart - Gonzaguinha).

Knowledge requires a curious presence of the subject in the face of the world. It requires transformative action on reality. It demands a constant search. It involves invention and reinvention.

Paulo Freire

SUMMARY

Executive Functions are among the most complex aspects of cognition and involve different components, and such complexity is also reflected in conceptual definitions, since there is still no consensus about their definition. Currently, however, several authors agree that executive functions are an integrated system comprising distinct domains such as flexibility, updating, inhibition, alternation, and planning.

The complexity of the construct reflects directly on the assessment of executive functions and the construction of appropriate tasks, since they do not measure only one skill. Additionally, several studies emphasize that executive functions develop throughout the child's growth, and the relationship with academic performance is well established in the literature. The present study is part of a larger project whose main objective is to develop and propose a battery that evaluates distinct domains of executive functions, so that it can be used with several participants of different age groups, from infancy to adulthood, and also by different intellectual and socioeconomic levels.

The objective of the present study was the development of the battery for children aged 6 to 12 years old. The study was developed in successive steps: a) consultation of current literature on the main paradigms for the assessment of executive functions for the adaptation/construction of tasks; b) pilot studies to analyze the adequacy of the battery; c) testing in a group of children with Attention Deficit Hyperactivity Disorder.

The results showed that the battery presents tasks that seem adequate to the purpose of assessment, but with others that still need changes. Future studies will determine that the tasks may thus be more specific in the domains and components of the executive functions. And yet, to be of public domain, not requiring the use of paid equipment or software, and that can be employed from the beginning of schooling to adulthood in people of different socioeconomic levels. In this way, neuropsychological assessment becomes more precise, sensitive and specific with regard to each executive domain, helping in the diagnostic process.

Key words: Executive functions, Neuropsychological Assessment, Children, Development.

Summary

CHAPTER 1

INTRODUCTION

Neuropsychology, especially Cognitive Neuropsychology, has studied Executive Functions, which comprise the set of skills, allowing the execution and development of voluntary actions oriented to goals. It involves cognitive and metacognitive processes, as well as emotional ones, enabling the individual to perceive and respond adaptively to stimuli them (ALFANO et al., 2008).

The executive functions consist of a set of skills which, in an integrated way, allow the individual to direct behaviours towards goals, evaluate the efficiency and adequacy of these behaviours, abandon inefficient strategies in favour of more efficient ones and, in this way, solve immediate, medium and long term problems.

The development of the executive functions of FE is an important adaptive landmark of the human species because it is related to several of our behaviours, such as the ability to deal with groups, to accept social rules, the ability to learn and imitate from observation, among others.

Besides the relationship with academic performance, FEs have been related to social and mental problems, such as Attention Deficit Hyperactivity Disorder (ADHD), Global Developmental Disorders, intellectual disability, disruptive behaviors and school dropout (HARTMAN et al., 2010, MAZZOCCO; KOVER, 2007, ARNOUDSE-MOENS et al., 2009).

Many are the difficulties found to accurately assess the FEs, whether in adults or children. This is because most of the instruments end up assessing more than one domain, and thus makes it difficult to accurately analyze the degree of impairment (MIYAKE et al., 2000).

Thus, these factors reveal the importance of studying new neuropsychological assessment procedures that may favour diagnostic accuracy. As highlighted by Carim et al., (2012) "Although executive functions are crucial for the effective functioning of the child at school and in society as a whole, there is a scarcity of appropriate instruments to assess this construct"

Thus, the present research is part of a larger study, whose title is "Effects of development and socioeconomic level in different domains of executive functions", whose main researchers are from the Department of Psychobiology- UNIFESP. The project aims to propose a battery of tests that assesses six distinct domains of EFs, which can be employed in subjects with different socioeconomic levels (from illiterate to people exposed to high levels of cognitive stimulation) and ages (to be employed with children from 6 years old to adults with 60 years old). Therefore, the battery does not include reading and writing stimuli, the tests will be in the public domain (not paid) and can be performed without the need for equipment or computer programs.

Based on the studies of Myake et al. (2000), who evidenced 3 specific domains: (1) switching between tasks or mental sets; (2) updating and monitoring representations in working memory; (3) inhibition of dominant or preponderant responses. Dual *task performance was* not related to any of the three domains, suggesting a separable ability from these three basic FEs studied.

Two other executive processes that were not assessed by Miyake et al. (2000) have also been defined as distinct cognitive domains: planning, that is, the ability to organise behaviour with respect to a specific goal that should be achieved through intermediate steps (OWEN, 1997, SHALLICE, 1982); and the efficiency of access to long-term memory (FISK; SHARP, 2004).

Specifically, the present study analyses the suitability of the proposed battery for the age group of children between 6 and 12 years old,

CHAPTER 2

LITERATURE REVIEW

2.1 EXECUTIVE DUTIES

As described by Carim (2012) human beings have an enormous capacity to seek challenges and accomplish their goals. To this end, it will use highly qualified cognitive functions called Executive Functions. The Executive Functions (EFs) allow us to organize behavior, meet immediate demands in favor of long-term goals.

Many are the doubts about the concept of Executive Functions (EFs), whose neuropsychological assessment has been a growing object of great interest of researchers (MIYAKE, 2000; HELEN et al., 2006; HAMDAN, PEREIRA, 2005; DIAS et al., 2010; VAN DER VEN et al., 2012).

The International Neuropsychological Society dictionary defines FEs as "the cognitive abilities required to perform complex goal-directed behaviours and the adaptive capacity to diverse environmental demands and changes" (LORING, 1999, p. 64).

Some authors consider them as the "conductor of the individual's mental symphony" (BARKLEY, 1998). For Ardila and Otrosky-Solis (1996), they are the various skills and cognitive abilities of the participant to engage in goal-oriented behavior, performing voluntary actions, independent, self-organized and directed to specific goals, i.e. cognitive skills involved in planning, initiation, follow-up and monitoring of complex behaviors directed to an end (HAMDAN; PEREIRA, 2009).

Lezak et al. (2004) describe that FEs are characterized by processes intrinsic to the ability to behave and respond in a way adapted to new situations and are composed of four components: volition, planning, purposeful behavior and effective performance. Volition refers to the ability to engage in purposeful behaviour with purpose and intention, involves memory functions, impulse control and sustained attention. Planning refers to the identification and organisation of elements and steps to reach such a goal and involves the capacity for flexibility and inhibition of responses.

Purposeful behaviour involves programmed tasks, ensuring effective performance, these components are constantly subjected to a monitoring system (DIAS, 2009).

Currently, FEs are referred to as an umbrella term because they cover several cognitive domains, called *hot* and *cool* (CHAN et al. 2008). Hot skills involve emotions, beliefs or desires, regulation of social behavior, decision-making, personal interpretation, and activities that contain reward or punishment (CHAN et al. 2008). Cold skills are those based mainly on logical reasoning, such as planning, sequencing, problem solving, cognitive flexibility, and the ability to deal with new situations (CHAN et al. 2008).

Due to the difficulties in defining FEs, several theoretical models have been proposed, although so far none has been unanimously accepted by researchers interested in the subject (CHAN et al., 2008; PACKWOOD et al., 2011; UEHARA et al., 2013):

a) Luria's Three Functional Units Theory (1981) - describes FEs as higher psychic functions that play an important role in the regulation of wakefulness and in the control of more complex goal-directed human activities. According to Luria, the brain is composed of three basic units which interact with each other.

b) Supervisory Attentional System (SAS), by Norman and Shalice (1986) - highlights the role of attention in active behaviour and distinguishes automatic actions (learned or habitual behaviours) from those requiring defined, controlled attentional resources (non-habitual behaviours, involves planning and decision-making).

c) Stuss and Benson's Tripartite Model (1986) - there are three systems acting in the monitoring of attention and FEs; two of them are concerned with the maintenance of the individual's state of alertness, while the third performs executive control (planning, selection of stimuli, of responses).

d) Duncan's Meta-Neglect Theory (DUNCAN et al., 2000) - emphasizes goal-oriented behavior, the role of the set of goals and subgoals in the proper functioning of behavior. Goals would be formulated, stored and checked in the mind in order to organize the appropriate behavioral response to the demands.

e) Somatic Markers, by Damasio (1996) - highlights the function of the frontal lobe in emotion and social behaviour, especially in decision making.

f) Executive Functions in Four Domains, by Lezak, Howieson and Loring (2004) - conceptualizes FEs as capacities that allow someone to maintain self-regulatory behavior, independently and intentionally. It proposes an explanatory model for FEs based on four components: volition, planning, premeditated behaviour and effective performance.

g) Baddeley's (1996, 2002) Working Memory Model - working memory consists of a limited attentional capacity system and two "slave" systems. The functions of the "central executive" include selective attention, coordination of two or more competing activities, *switching attention* and retrieval of information from long-term memory (BADDELEY, 2000).

h) Barkley's (1997) Executive Self-Regulatory Model - self-regulation is a major component of executive functioning. "It is any response given with the intention of altering a subsequent response to another event, and thereby altering the probability of a subsequent consequence related to that event" (BARKLEY, 1997). It incorporates most of the key components of FEs: goal-directed behaviour, devising plans to achieve future goals, use of rules, plans and impulse control. Prerequisite: behavioural inhibition.

i) Problem Solving Framework, by Zelazo et al. (1997) - a group of sub-functions and a problem solving network that describe the distinct phases of executive functioning. This model involves several executive processes, or sub-functions, that work together to achieve a given goal: representation of the problem; planning and selection of the most appropriate actions for greater efficiency; execution and evaluation which involves the detection of errors and their correction;

j) Independent and Interrelated Domains, by Miyake et al. (2000); St Clair- Thompson and Gathercole (2006); and Van der Ven et al. (2012) - treats executive functions as an integrated system comprising distinct domains. According to the model, although independent, they interact with each other. This theoretical model, which is the basis of the present study, will be more extensively addressed below.

Thus, this diversity of theoretical models attempts to explain executive functioning and its components and, despite the lack of agreement, such a movement tends to motivate the development of sensitive and specific measures, further raising the desire for knowledge about the functioning and deficits of executive functions.

2.2 Neurobiological Basis of Executive Functions

Activities related to FEs are performed by the frontal lobes of the brain (BURGESS; ALDERMAN, 2004), more specifically the prefrontal cortex (DUNCAN, 2001), which establishes connections with cortical regions (sensory and motor) and also with subcortical regions (basal nuclei and thalamus) (GAZZANIGA et al., 2006).

Initially it was believed that there was no frontal lobe activity during childhood, but this assumption was soon refuted (ANDERSON, 2001, ANDERSON et al. 2001). Several processes (arborization, millienization and synaptogenesis) are described with a hierarchical development during childhood and adolescence and the development of FEs are aligned with these neurophysiological changes in the prefrontal cortex (CPF) (ANDERSON, 2002).

There are distinct regions of the CPF that are responsible for processing different FEs (cold or hot) (BECHARA et al., 1998). Cold skills activate to a greater extent dorsolateral regions that establish connections with motor areas, basal ganglia, cingulate gyrus, sensory association areas and parietal cortex. Hot skills involve orbitofrontal and ventromedial regions, establishing connections with the amygdala, hippocampus, visual and temporal association areas, and are related to the integration of emotional, motivational and social information (WOOD; GRAFMAN, 2003, CHAN et al. 2008), which explains their impact on decision making.

In both phylogenesis and ontogeny, the frontal lobe is the region that develops later and in stages, starting with the primary motor and sensory areas, moving to adjacent secondary areas and ending with the parietal, temporal and prefrontal association areas (STUSS; ANDERSON 2004).

At birth, the development of the frontal cortex is not yet complete and its maturation will occur during adolescence to early adulthood, which results in a limited capacity of FEs in children and adolescents, facing the demands imposed by the environment (ANDERSON, 2001).

Many researchers cite the prolonged development of the prefrontal cortex as the main mechanism for FE maturation (CONKLIN; LUCIANA; HOOPER; YARGER, 2007).

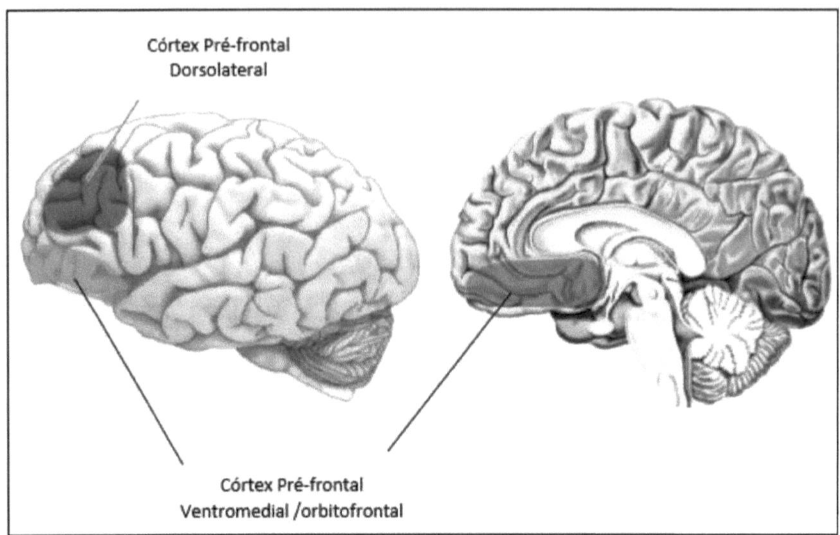

Figure 1 - Dorsolateral ventromedial/orbitofrontal prefrontal cortex in left lateral view and in sagittal cut

2.3 Development of Executive Functions

Several studies point out that FEs develop throughout the child's growth, and indicate periods with important alterations during this process (ROMINE; REYNOLDS, 2005, WELSH; PENNIGNTON, 1988).

Despite the vast literature on FEs in children, there is little information about the processes they go through from one level to another in relation to the development of FEs from infancy to adolescence (BEST; MILLER, 2010).

In this regard, Piaget (1954) described that before 7 or 8 months of age, children do not look for objects hidden under a lid, because, for them, these ceased to exist (PIAGET, 1954). As early as 9 months, they develop simple forms of mental representation, such as searching for a hidden reward in a goal-directed manner (Piaget's A not B paradigm) (BERK, 2006). This indicates that children under the age of 1 year generally have difficulty maintaining memory for long periods of time (WELSH; FRIEDMAN; SPIEKER, 2006). On the other hand, after the age of 12 months, they already tend to search for hidden objects in various places successfully (PIAGET, 1954).

From 2 years on, changes occur and the child becomes an intentional being that can follow verbal rules (keep verbal rules in mind and use them to guide his/her behavior) and its evolution continues until 4 years of age (ZELAZO; MUELLER, 2002). Between 6 and 8 years of age, it is possible to notice the ability to self-regulate behavior, behaviors, and improvements in the ability to set goals and anticipate events without depending on external instructions, since they still show simple forms of organized search and planning, with improvement of these abilities (PAGE, 1985).

In this sense, Romine and Reynolds (2005) gathered seven studies and examined age-related changes in executive skills in children and adolescents aged 5 to 17 years. The research revealed that the main increases occur between the ages of 5 to 8 years in the skills of planning, fluency, inhibition and alternation.

Also Best and Miller (2010), in a review, analysed some studies in an attempt to outline the general developmental trajectories of FEs, in preschool and school-age children, or school-age children and adolescents. The authors observed that, although FEs emerge during the first years of life, they continue to strengthen significantly throughout childhood and adolescence, although some components vary in their developmental trajectories, as is the case of the skill of inhibition, which shows improvement at ages 3 to 6; stabilizes in the early school years; does not show significant changes between ages 7 to 13; and shows progressive improvement again from ages 15 to 21.

These authors also showed a linear increase in the development of operational memory for simple tasks at ages 4 to 14 and a levelling off between ages 14 and 15. Switching is another skill that shows significant improvement with age. Children aged 3 to 4 years can successfully switch between two simple response sets where rules are placed in a context and inhibition demands are reduced. The main improvement occurs in the 5 to 6 year old range and reaches adult-like levels by age 15.

Confirming this process, Juhani et al. (2010) analysed the development of various stages in different components of FEs progressing from childhood to adolescence. The study showed that in most tasks, participants' performance improved with age - nine of the 14 executive measures used correlated significantly with age. According to Zelazo and Mueller (2002), around the age of 12, children can achieve performance similar to that of an adult in some FE tasks.

Changes in executive functioning throughout life are characteristics of human development. In general, it can be expressed as an inverted U-shaped curve, in which after the fourth decade of life a discrete decline from a previous level of functioning is expected (ZELAZO; CRAIK; BOOT, 2004).

2.4 Neuropsychological Assessment of FEs

In clinical practice, neuropsychological assessment is used to verify the presence of deficits or cognitive or behavioral impairments that may be of neurological origin (MIRANDA; MUSZKAT, 2004). In them, FEs are among the set of skills that stand out the most and, therefore, should be one of the cognitive domains subject to a comprehensive study (NATALE; TEODORO; HAASE, 2008).

There is therefore a need to establish evaluation measures that are sensitive to the changes inherent to age and that quantify the alterations, allowing the difference and the limit between normal and pathological to be reliably defined.

For the assessment of FEs, there are numerous scales, tests and batteries (SPREEN, 1998). Among the most used are the Winsconsin Letter Test (WCST), the Stroop, the Trail Making Test (TMT) and the Tower of Hanoi. For some researchers, the WCST has been considered the gold standard for the evaluation of FEs (HAMDAN; PEREIRA, 2009).

However, in other studies, the WCST, the Stroop and the TMT showed no alterations in patients with executive deficits (RABIN et al., 2005, BURGESS et al., 2006). Miyake et al. (2000) also showed that performance in classic executive tasks, such as the WCST and the Tower of Hanoi, reflects the use of the three executive domains, that is, inhibition, alternation and updating, constituting measures that are not representative of distinct types of FEs.

Another example of evaluation is the ecological battery Behavioral Assessment of the Dysexecutive Syndrome (BADS) (WILSON et al., 1996), which includes different tests that evaluate the capacity of organization and planning.

There are other FE batteries, such as: Cambridge Neuropsychological Test Automated Batteries (Cantab) (FRAY et al., 1996), which assesses learning, operational memory and planning, alternation, speed, sustained attention and fluid intelligence; Delis-Kaplan Executive Function System (D-Kefs) (DELIS et al., 2001), which assesses alternation, inhibition, planning, response generation, deductive reasoning and concept formation.

The complexity of the construct, referred to above, reflects directly on the assessment of FEs and the construction of appropriate tasks, since they do not measure only one skill and most often assess non-executive processes.

Still in this direction, several researchers address relevant issues about FEs and their assessment, that is, the instruments used and whether they are coordinated by a single function or act as independent modules, further demonstrating the need to define this construct (VAN DER VEN et al., 2012, MIYAKE et al., 2000).

Recent evidence suggests that these functions do not constitute a single entity, presenting several sub-processes or domains (BADDELEY; DELLA SALA, 1996, MIYAKE et al., 2000, SMITH; JONIDES, 1999).

In a study with adults (37 university students), Miyake et al. (2000) confirmed the existence of these three main components of FEs and analysed the individual differences and correlations of these functions in the performance of complex tasks (alternation, inhibition and updating).

The authors used confirmatory factor analysis, indicating that the three FEs are moderately correlated with each other, but are also clearly separable, showing the different contributions of FEs in certain tasks, which demonstrates the unity and diversity of executive processes.

Furthermore, they indicated that *dual-task* performance, or the ability to perform two tasks at the same time, was not related to any of the three domains cited (switching, updating and inhibition), suggesting that multitasking may be a separable skill from the three basic FEs studied.

Planning ability has also been defined as distinct cognitive domains, that is, the ability to organize behavior in relation to a specific objective which must be reached by means of intermediate steps (OWEN, 1997, SHALLICE, 1982). Another executive domain proposed is the efficiency of *access to long-term memory* (FISK; SHARP, 2004), processes which were not assessed by Miyake et al. (2000).

Another research, which also used confirmatory factor analysis, investigated the factor structure of FEs (inhibition, alternating and updating), their longitudinal development related to maths skills. It was evaluated 211 children, aged between 7 and 8 years old, and the result indicated that inhibition and alternation could not be distinguished. Updating was considered a separate factor, and its development was strongly related to that of mathematics learning (VAN DER VEN et al., 2012).

St Clair-Thompson and Gathercole (2006) assessed 51 children, aged 11 and 12, comparing school level and the FEs of updating, inhibition, alternation, including verbal and visuospatial operational memory. Their results differ from those of Miyake et al. (2000) in that they do not identify a third FE domain, that of alternation.

This disparity may reflect a fundamental difference in the organisation of FE between children and adults. Another finding in the results of St Clair-Thompson and Gathercole (2006) highlights that inhibition may not be considered a distinct component because the tasks for this domain are not "pure" and their performance requires, albeit minimal, working memory skills.

Also relevant finding was that verbal and visuospatial measures of working memory are associated with updating skills but not with inhibitory control processes, a further finding that reinforces Miyake et al.'s (2000) findings of strong and specific links between updating and verbal working memory measure.

The subdivision of FE is also reported in studies of patients with injuries, by means of neuroimaging, which assess the neural substrates activated during the performance of different tasks, in general in normal subjects (JONIDES et al., 1997, COLLETTE et al., 2005)

Thus, there appear to be six distinct domains of FEs: alternation, inhibition, updating, dual tasking, planning and access to long-term memory (Table 1), which are currently postulated in the literature.

Table 1 - Domains of FEs

Domain	Definition
Alternation	Ability to shift attention between different sub-tasks or disengage from a strategy or stimulus that is not appropriate at the moment (BARKLEY, 1997, 1998).
Inhibition	Ability to block irrelevant stimuli (HUGHES et al., 2004), inhibit a predominant or automatic response; interrupt an ongoing response when it is no longer appropriate; or resist distracting interference (MIYAKE et al., 2000).
Update	Capacity to evaluate information, review the existing content in the operational memory and when necessary eliminate what is no longer relevant, incorporating more recent relevant information. It is closely related to working memory (LEHTO, 1996, MORRIS; JONES, 1990).
Dual task	Ability to perform two tasks at the same time. (FERNANDES; MOSCOVITCH, 2000, NAVEH-BENJAMIN; CRAIK; GUEZ; KREUGER, 2005).
Planning	Ability to organise behaviour with respect to a specific goal, which is to be achieved through intermediate steps (OWEN, 1997, SHALLICE, 1982).
Access to long-term memory	Ability to temporarily activate long-term memory (BADDELEY, 1996, FISK; SHARP, 2004).

Source: Own elaboration.

Data on heritability demonstrate that the unity and diversity of FEs are almost entirely genetic in origin. Friedman et al. (2008) indicated that FEs are correlated and separable by being highly influenced by genetic factors (99%) general and unique to particular FEs, which places them among the most heritable psychological traits.

Corroborating such hypotheses, the following domains of FEs are indicated to have a genetic basis: updating, inhibition and alternation, which are stable throughout the development of individuals and have a range of relationships with regulatory behaviours, with important implications both clinically and socially (MIYAKE; FRIEDMAN, 2012).

Implications such as these may stem from deficiencies in FEs, which have been documented in school-aged children (CLARK, PRITCHARD, WOODWARD, 2010; DIAS et al., 2010; TILMAN et al., 2013); however, compared to adult studies, the amount of research with children is scarce, yet some researchers have important findings.

Tamnes et al. (2010) conducted a cross-sectional study, with 98 children and adolescents (8-19 years old), who were tested with six tasks to assess FEs and one of the results suggests improvement in the performance of some tasks according to age.

A not very recent work, (EPSY et al., 2001), adapted some tasks and evaluated 98 preschool children, relating performance on FEs tasks with ages ranging from 26 to 66 months, with the intention of delineating distinct cognitive profiles among preschool children with the various neurological developmental disorders. The results indicated that performance on FEs tasks was largely related to age group, but not to gender or intelligence, thus suggesting that the tasks were age-sensitive.

The knowledge of tests and of a new battery that makes use of simple tasks and that assesses more specific components of the Fes, in order to obtain more "pure" measures, as suggested by Myake et al. (2000), may help professionals and caregivers who work in the diagnosis and treatment of individuals with various neurodevelopmental disorders and, specifically, the Attention Deficit Hyperactivity Disorder (ADHD), as well as in the discrimination of possible deficits of a child or adolescent with ADHD, since it allows establishing a correlation with functional impairments, in addition to helping in the measures to be taken regarding rehabilitation plans.

CHAPTER 3

OBJECTIVES

3.1 General Objectives

To develop and propose a battery of executive functions that evaluates the distinct domains of executive functions, based on current paradigms, which can be used with different participants, regardless of age and education.

3.2 Specific Objectives

1. To develop an executive functions battery for children aged 6 to 12 years old.
2. To analyse the adequacy of the battery, as to instructions, floor and ceiling effects in children with typical development.
3. Analysis of appropriateness in a clinical group of children with Attention Deficit Hyperactivity Disorder.

CHAPTER 4

GENERAL METHOD

As initially mentioned, the present study is part of a large project entitled "Effects of development and socioeconomic level on different executive function domains" (CNPq process 443195/2014-5, under analysis). The aim is to propose a battery of tests to assess different executive skills that can be employed in people of different ages and socioeconomic levels, as well as to study the development of 6 differentiable executive domains and whether they are susceptible to socioeconomic level and pubertal stage. The coordinator of the project is Prof. Dr. Sabine Pompéia; vice coordinator Prof Dra Monica C Miranda; Raquel Luna Antônio and Gislaine Zanin as associated researchers.

Only the steps and stages related to the objective of the present study will be described below. All procedures of the present study were approved by the Research Ethics Committee of Unifesp, by means of protocol 296.837 (main project) and protocol 186.770 (present master's project). (Appendix E)

3.3 Stage I: Development of the initial battery

3.3.1 Theoretical assumptions and pertinent literature

The extensive consultation to the current literature, presented above, was the basis for the proposal of the first version of the FE battery. Thus, based on the experimental paradigms of the studies of Miyake et al. (2000), Morris and Jones (1990), Towse and Neil (1998), Ho et al. (2004) and Van der Ven (2012), mainly, the process of building the battery was started.

The subtests of each domain evaluated are described below, as well as the data from the pertinent scientific literature, effectively used for each of the tasks initially developed.

□ **DOMAIN *SHIFTING***

□ ***Colour-shape***

This subtest was adapted from Miyake et al. (2004) and Friedman et al. (2008). Participants are shown a symbol-track (rainbow or tic-tac-toe) on a blue or orange geometric figure (a circle or square).

The instructions inform that when the symbol-track appears, the participant must always say the shape of the figure; and in the presence of the symbol rainbow track (block 2), he must always say the colour of the figure (Figure 2).

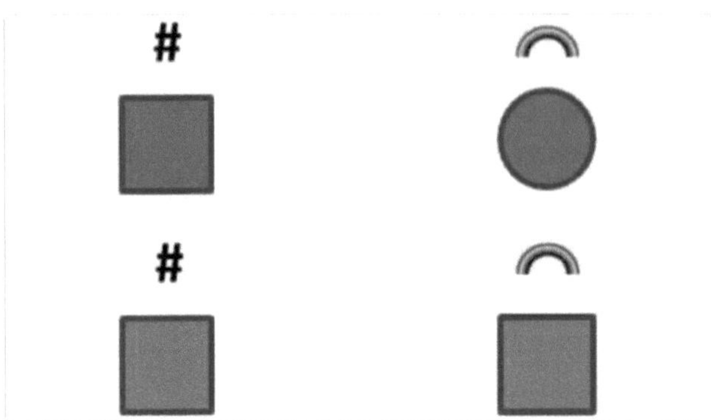

Figure 2– Colour-form subtest (blocks 1 and 2)
Source: Prepared by the project researchers.

In both blocks, 20 stimuli are presented. Half of the participants are asked to say the shape of the figure in the first block and the colour in the second; the other half should do the opposite, in order to avoid any order bias.

In the third block (Figure 3), the participant must alternate the response - colour or shape of the figure - according to the symbol-track. The clues alternate randomly, so that the ability to alternate the response quickly is an executive measure (response completion time and errors). The alternation task is composed of three phases. At the end, the cost of errors and time in performing the test is verified. From that, the summed performance during the performance of the two tasks without alternation in the two control blocks is subtracted (alternation cost).

Figura 3 - Colour-form subtest (alternation)
Source: Prepared by the researchers of the large project

□ **Category *switch***

Adapted from Mayr and Kliegl (2000) and Friedman et al. (2008), in each trial of this task, participants are presented with an image to be classified as alive or not alive, or big or small - for the purpose of comparison, they are instructed to think about whether it is bigger or smaller than a football.

The 16 images, adapted from words used by Mayr and Kliegl (2000), are based on studies in Brazil (POMPÉIA et al., 2001): four refer to large inanimate objects (mountain, house, Ferris wheel and bridge); four to small inanimate objects (nail, button, key and nail); four to large living objects (dinosaur, whale, giraffe and gorilla); and four to small living objects (butterfly, ladybird, ant and snail). Thus, each word can be classified in terms of two independent semantic dimensions: size and being/not being alive.

The images are presented on A4 sheet, and the track symbols (heart for the "to be or not to be alive" category and cross with arrows for the "big or small" category) appear above them (Figure 4).

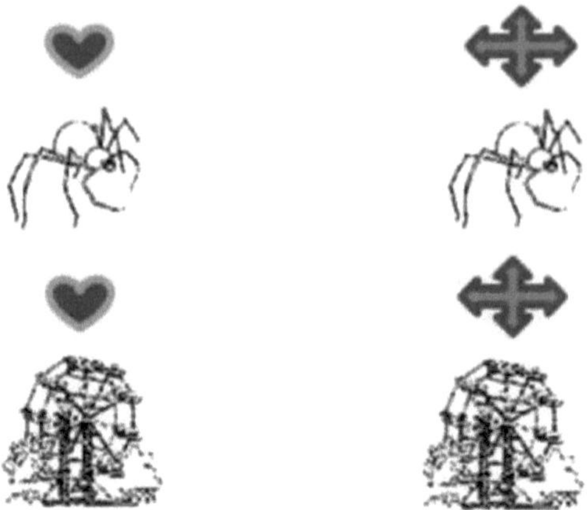

Figura 4 - Subtest alternating categories (blocks 1 and 2)
Source: Prepared by the researchers of the broad project.

In the first block, control, the participants must say only if the objects are alive or not alive. In the second block, also control, they must point whether the objects are large or small. In both blocks, 20 stimuli are presented each.

In order to avoid any order bias, half of the participants are asked to say whether, in the first block, the object is alive or not alive, and in the second block whether it is big or small.

In the third block, participants should alternate the response - living or non-living object, large or small - sequentially, during 40 stimuli: in one stimulus they should say whether it is living or non-living, and in the next, whether it is large or small, and so on (Figure 5). The ability to alternate the response quickly is an executive measure (response completion time and errors). From this, the sum of the performance during the performance of the two tasks without alternation in the two control blocks is subtracted (alternation cost).

Figura 5 - Subtest alternating categories (alternation)
Source: Prepared by the researchers of the broad project.

□ *INHIBITION* DOMAIN

□ Numerical stroop (*the stroop-numerical test*)

In the adaptation of the task by Fournier-Vicente et al. (2008), parts A and C of the original test were used, presented successively. The cards (A4 size) contain one hundred stimuli, with ten lines of ten stimuli each.

In the first part, blocks of two, three, four or five arrows (e.g., TUD are presented to the participants, who are instructed to say aloud the number of arrows in each block as quickly as possible, leaving no uncorrected errors (Figure 6).

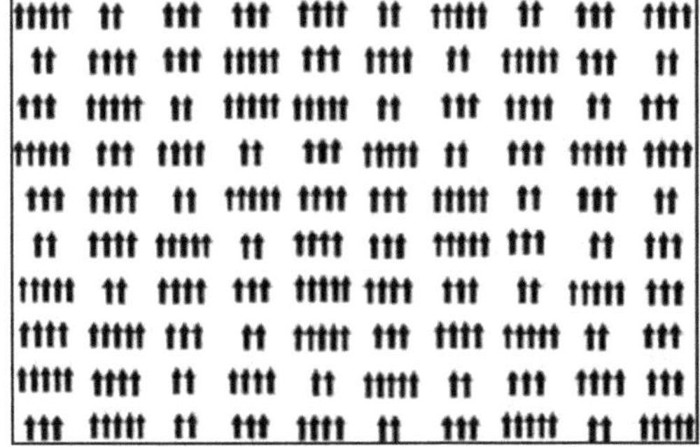

Figura 6 - Numerical Stroop subtest (part 1)
Source: Prepared by the researchers of the broad project.

In the second part, the stimuli consist of digits presented in blocks of two, three, four or five digits each, all from one block with the same value (e.g., 2222), but with no correspondence between the number of digits in a block and the value of digits contained

(Figure 7).

22222	44	33333	55	222	33	5555	44	22222	3333
33	2222	44444	555	22222	444	55	3333	55	22222
22222	555	3333	44	33333	555	2222	444	33333	5555
3333	444	22222	55	44444	33	222	5555	44	222
5555	222	44444	33	22222	444	5555	33333	2222	44
444	33333	555	222	33	555	44444	2222	555	3333
55	2222	44444	3333	2222	444	55	3333	5555	222
2222	33	55	444	5555	222	44	33333	5555	33
33333	55	2222	444	555	44	33333	222	33	44444
444	22222	3333	55	222	5555	3333	44	555	44444

Figura 7 - Numerical Stroop subtest (part 2)
Source: Prepared by the researchers of the broad project.

Participants are instructed to say aloud the number of digits in each block as quickly as possible and without errors (e.g., for 2222 they should say that there are four digits). For correction, the performance of the second part - in which the number of digits and the value in each block are incongruent - is subtracted from that of the first part, which measures stimulus counting speed and verbalization without incongruence, generating the dependent measure (inhibition cost).

- ***Go/No-go***

In this task, adapted from Casey et al. (1997), the stimuli, presented sequentially, are numbers from zero to nine. The volunteer must answer yes to any number except 4, to which he/she must remain quiet.

In the first block, 100% of the stimuli (i.e. 20) are targeted (all different from 4 - the volunteer should answer yes); in the second block, from a total of 40 stimuli, 50% are targeted (numbers different from 4 - the volunteer answers yes) and 50% are non-targeted (number 4 - the volunteer does not answer) (Figure 8).

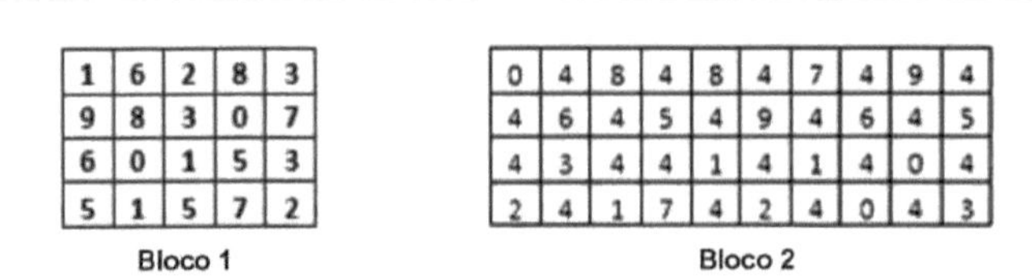

1	6	2	8	3
9	8	3	0	7
6	0	1	5	3
5	1	5	7	2

Bloco 1

0	4	8	4	8	4	7	4	9	4
4	6	4	5	4	9	4	6	4	5
4	3	4	4	1	4	1	4	0	4
2	4	1	7	4	2	4	0	4	3

Bloco 2

Figure 8 - *Go/No-go* subtest (blocks 1 and 2)
Source: Prepared by the researchers of the broad project.

In the third and final block (Figure 9), 80% of the stimuli are target (any number but 4) and 20% are non-target (the number 4), and 200 are presented. The total execution time of the first block is multiplied by 6, resulting in 120 target stimuli. The total execution time of the second block is multiplied by 2, resulting in 40 target stimuli and 40 non-target stimuli (number 4). Both results are added and we have the task control time, of low demand regarding inhibition, in which 80% of the stimuli are target and 20% are non-target, keeping the proportion of the third block. From the execution time of the third block, the control time is excluded, determining the dependent measure. The number of errors, false alarms and absence of response are still accounted.

1	5	4	8	3	9	4	2	5	7
0	4	3	1	5	7	2	6	4	8
3	5	7	2	4	3	1	5	9	4
5	7	6	4	3	1	4	2	6	0
8	4	6	1	9	3	0	4	2	5
0	7	4	9	8	4	1	6	3	7
4	0	7	8	0	6	4	5	2	1
6	8	3	4	5	9	7	2	4	0
2	6	8	0	3	4	5	7	1	4
9	1	4	6	5	0	2	4	3	8

7	4	1	0	4	8	3	2	9	5
4	8	3	9	7	2	0	6	4	1
5	2	0	4	9	4	6	8	1	7
3	8	4	5	0	5	1	9	7	4
4	2	5	0	8	3	4	7	9	6
1	8	0	9	4	6	2	4	5	3
6	4	9	7	1	4	3	5	8	2
0	7	2	4	6	3	1	8	4	5
3	1	4	5	7	2	6	0	9	4
4	8	2	5	9	0	4	3	6	1

Figure 9. *Go/No-go* subtest (block 3)
Source: Prepared by the researchers of the broad project.

□ DOMAIN ***UPDATING***

○ Random Number Generation (*RNG)*

Adapted from Towse and Mclachlan (1999), in this task, participants must randomly generate 70 digits using digits 1 to 10, every 2.5 seconds (ages 6 to 10 years). Participants over the age of 10 must randomly generate one hundred numbers at a rate of one digit every 1.5 s. For both children and adults, the sequence of numbers generated is noted and scores are assigned according to the criteria suggested by Towse and Neil (1998) and Ho et al. (2004).

Randomness indices are calculated by the RgCalc *software* - whose distribution is free, via the Internet (<http://www.pc.rhbnc.ac.uk/cdrg/rgcpage.html>) -, which reflect specific subtypes of FEs (JAHANSHAHI et al., 2006, MIYAKE et al., 2000). An articulation check was performed, asking the participant to count from 1 to 50 (or 1 to 10, five times) as fast as they could.

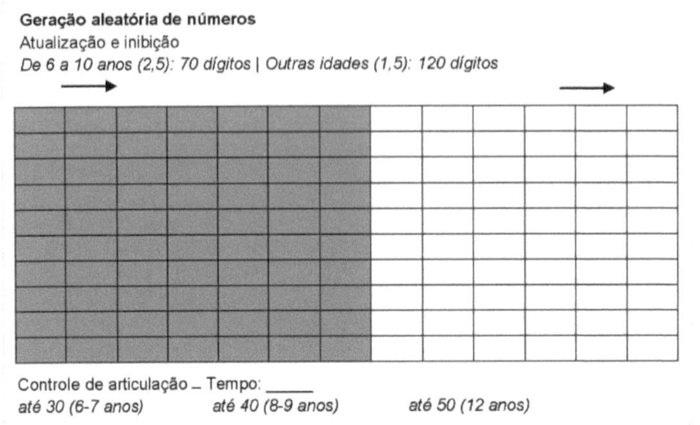

Figura 10 - Random number generation subtest (part of the application protocol)

Source: Prepared by the researchers of the broad project.

- **Updating words**

In this subtest, adapted from Morris and Jones' *letter memory* paradigm (1990), several words with mono- and disyllable names are presented in sequence. The task is to recall the last three. To ensure the continuous updating task, participants should say aloud the last three words every time, then add the most recent figure, to finally say the new sequence of three words aloud. For example, if the number of words presented is 6 ("bell, candle, moon, wheel, cake, flower"), the participant should say, as the words are heard: "bell... bell; candle... bell, candle; moon... bell, candle, moon; wheel... candle, moon, wheel; cake... moon, wheel, cake; flower... wheel, cake, flower"; and remember at the end of the task to "wheel, cake, flower".

The number of words presented (4, 6, 8 or 9) is varied across trials. The amount of stimuli is increased to make it more difficult to identify the ease or difficulty of the task. As in the Friedman et al. (2008) study, participants are instructed to recall the words in order and say "blank" if they cannot remember any. Responses are scored as correct even when the words are not said in the correct order. After practicing three trials (one of each quantity), participants complete 12 sequences (four of each total quantity of words).

- ***Keep track***

Adapted from Miyake et al. (2000) and Yntema (1967), in this task, the examinee views a series of images, each belonging to one of the following three categories: kitchen items, toys and clothes/accessories. Participants are first introduced to the target categories and the exemplars of each to ensure that they know which category each picture belongs to. Examples of categories: (a) kitchen items: cup, cooker, spoon, cup, glass, pan, knife, bottle and refrigerator; (b) toys: ball, kite, spinning top, doll, clown, bladder, dominoes and roller skates; (c) clothes and accessories: shoe, glove, skirt, glasses, sock, tie, watch and ring (Figures 11, 12 and 13). One image is presented at a time and the participant is asked to name each one together with the applicator.

Figura 11 - Category kitchen items
Source: Prepared by the researchers of the broad project.

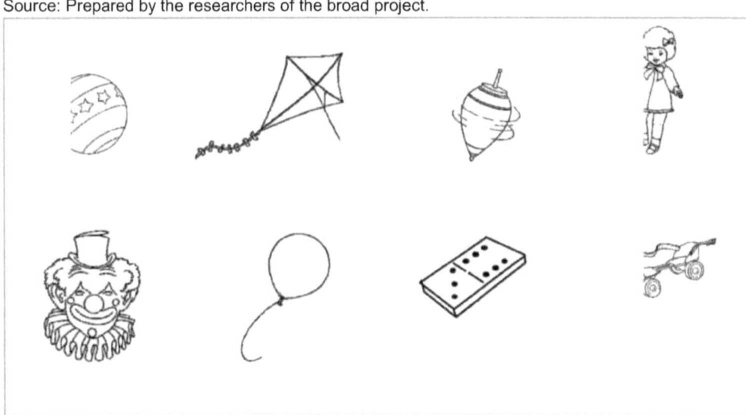

Figure 12 - Toy category
Source: Prepared by the researchers of the broad project.

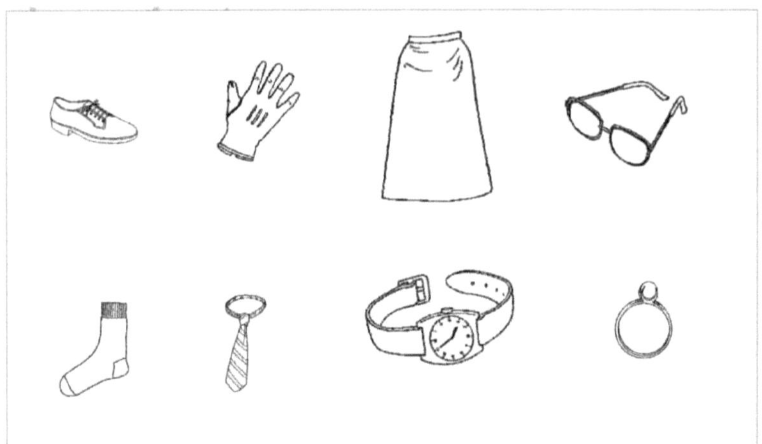

Figure 13 - Clothing and accessories category
Source: Prepared by the researchers of the broad project.

After the presentation of the categories, the task consists of eight lists of 15 images, including one, two or three copies of each of the three possible categories, which are presented serially and in random order (1,500 milliseconds (ms) each for adults and 3,000 ms each for children). Other images which are not part of any of the categories presented complete the lists. At the end of the presentation, the examinee should recall the last item from the designated categories.

The number of categories to be remembered progressively increases. One category is of only one target stimulus (memory control), and in the other two, updating progresses in 2-3-4-5 item difficulty. Three trials serve as training (training 1 - bottle, skates, glove, banana, globe, hat, cup, mouth, pineapple and pen - correct answer: skates, cup, hat); then participants perform two trials with each target category, presented in random order, recalling a total of 24 words (Figure 14). The proportion of correctly recalled words is a dependent measure.

Figure 14. Training 1
Source: Prepared by the project researchers.

□ PLANNING DOMAIN

○ Park Task

Planning task adapted from Wilson et al. (1996) and the BADS ecological battery. In it, a map of a park is presented (Figure 15), with pictures instead of place names, and a set of schematic instructions related to the places that participants have to visit and with rules they must obey (Figures 16 and 17).

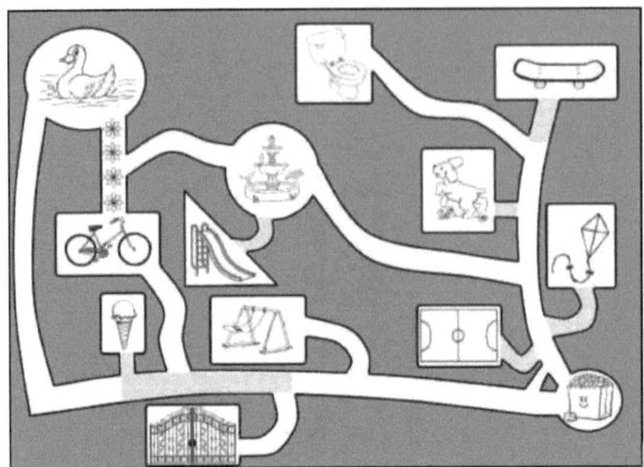

Figure 15 - Subtest of the park task (map)
Source: Prepared by the researchers of the broad project.

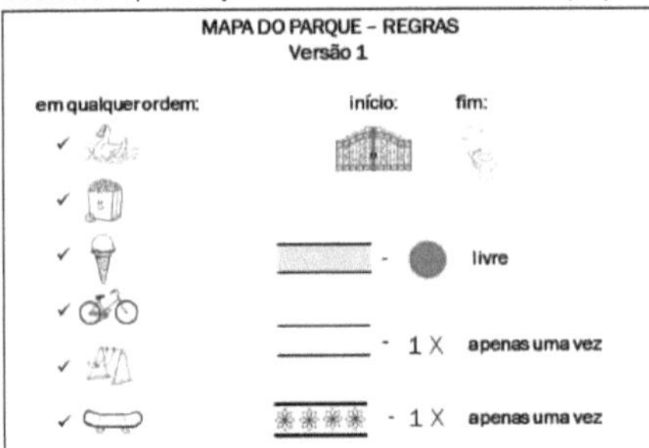

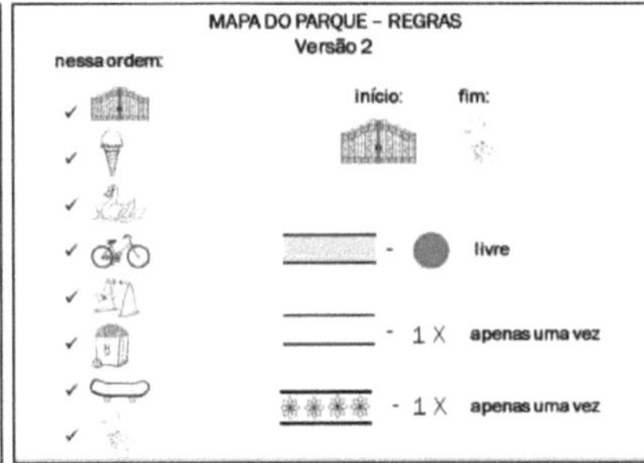

Figure 17- Subtest park task (1ª rule)
Figure 16 _ Subtest of the park task (2ª rule)
Source: Prepared by the researchers of the large project

There are two blocks with identical targets, which involve a visit to six, out of 12 possible locations. The first block consists of a high cognitive demand version, since there are paths that can be passed only once, and therefore to visit the required places, the planning skills of the participants are tested. In the second block, or low demand version, the participant is guided to simply follow instructions to achieve set goals. The score is given by the difference in time to perform the high demand task and the low demand task (ALLAIN et al., 2005).

These were the subtests initially selected to compose the

FEs. Regarding the stimuli used in the tests described above, we chose not to include written material or material that might require reading ability. All verbal stimuli, of high frequency in Portuguese language, were selected from the study by Pinheiro (1996). No stimulus or category of stimuli was repeated in different tests.

4.1.2 Adapting and constructing the instructions for children and adolescents

After the selection of the tasks presented, we proceeded with the adaptation and construction of the instructions for each task. Taking into account the characteristic of this battery that it can be used with different participants, regardless of intellectual level and age, the instructions were adapted from the original instruments. Those that were not available were created.

The aim of this phase of the research was to make the understanding of the instructions as clear as possible for participants under or over 10 years of age. The instructions are presented below.

GENERAL INSTRUCTION - PILOT

We are assessing a **test set** that has **10 tasks**. These tasks assess 'executive functions', which are functions in our brain that help us to organise, decide and plan our actions.

You are going to do these 10 tests so we know if our battery, our test set, is good, **if the people doing these tests can understand our instructions** and what they should do in each of the tests. **We still don't know** if these tests are good, and so we **need your help to** test them. So don't worry because **there is no right or wrong answer**, if you "do badly" on the test it means that we haven't prepared you properly.

Among the tests we will be doing, you will have to say words quickly, remember numbers, remember pictures, find paths on a park map, copy pictures, respond when you hear some kind of whistle, and the like.

You will find some of these tasks easy, while others may be more difficult. Some people do not answer all the questions correctly or do not complete all the items. I urge you to try your best to solve all the items.

Do you have any questions? Can we get started?

- **Colorform**
  - 1° block: You will now see a shape, either a circle or a square. **This symbol** (point to the "tic-tac-toe" symbol) will appear on top of the picture, indicating that you must say **the shape of the picture: circle or square**, as **quickly as you can**. You will repeat this task a few times. Any questions? Can we start?

- 2nd block: You will see the same pictures, but now the symbol that appears on top of the picture is **this rainbow** (point to the rainbow), so you have to say **the colour of the picture: blue or pink**, **as fast as you can**. You will repeat this task a few times. Any questions? Can we start?

- 3rd block: This time you will see the **same pictures** as before, but this time **the symbol** on top of the picture **will vary**: when the **tic-tac-toe symbol** appears, you must say the **shape of the picture,** circle or square; when the **rainbow** appears, you must say the **colour of the picture,** blue or pink. Do this as quickly as you can. You will repeat this task a few times. Any questions? Can we start?

□ **Alternation of categories**

- 1st block: You will see some images. Say, as **quickly as you can**, whether the picture **describes something living or non-living/dead**. You will repeat this task a few times. Any questions? Can we start? Let's have a practice run.

- 2º block: you will see the **same pictures**, but now you must say as **quickly as you can** whether the picture represents something **big or small**. To find out whether the object is big or small, think about the size of **each picture compared to a football**: if it is bigger you can say 'big', if it is smaller you can say 'small'. You will repeat this task a few times. Any questions? Can we begin?

- 3rd block: You will see the **same pictures you** saw before, but this time **you must alternate between the** classifications, between saying if the picture is **something alive or not alive (when it is not alive you can say - dead** or if it is **big or small**. That is, in the first picture if you say whether it is "alive or not alive", in the next picture you must say whether it is "big or small", and so on. You will have to **remember** which classification you used in the previous picture to be able to classify the next picture. You will repeat this task a few times. Any questions? Can we get started?

□ **Numerical stroop**

□ **Item 1**

Instruction: You are going to say out loud the number of arrows that appear in each block as quickly as you can, without making any mistakes. For example, when you see this block TfT, say 4.

We are now going to have a training session.

Itens de treino

↑↑↑↑ ↑↑ ↑↑↑ ↑↑↑↑ ↑↑↑ ↑↑↑↑↑ ↑↑ ↑↑↑↑↑ ↑↑↑ ↑↑↑

That was a practice run, now do the same thing you did before. Say the number of arrows that appear in each block. Can we start? Now!

□ **Item 2**

Instructions: Now you are going to see several numbers. Say out loud how many times the number appears, do not say which number you are seeing, just say how many times it appears. For example, when you see 2222 say "four" and not "two". Do this as quickly as you can without making any mistakes.

We are now going to have a training session.

Itens de treino

2222 33 444 5555 222 3333 55 44444 555 444

That was a practice run, now do the same thing you did before. Say how many times the number appears in each block. Can we start? Now!

□ ***Go/No-go***

In this task you will hear some numbers and I need you to say YES to every number you hear, EXCEPT number 4. When you hear 4, do not say anything, be quiet. Got it? So let's have a practice run. (*Use the same instruction for all three blocks of the task*)

□ **Random number generation**

For under 10s

(*Show the dice.*) Do you know what it is? It looks like a dice, but a bit different because it has 10 sides instead of 6. I'm going to ask you to roll the dice a few times and tell me what number comes up. (*Let the child roll the dice*)

Can you tell which number will appear when you roll the dice again? No, we don't, because any number can come up. Well, now I want you to pretend that it's a die of this type and that every time you hear a beep you imagine you've thrown the die and tell me what number has come up. I want you to try to say the numbers as they would appear if you were rolling the die, or the numbers all messed up and out of order. So, if you say the number 1 and then you say 2, then 3, 4, 5, 6, 7, 8, 9, 10, is that how the numbers would look if you were rolling the dice? That's not quite right, because the numbers all look out of order when you roll the dice. I want you to try to say the numbers in the messiest order possible, like a die. So, you have to say a number every time you hear a beep, OK? Let's practice? (*At the end of the test the children will be congratulated for being "great dice". You will then be asked whether the child found the test easy or difficult*).

For over 10s

You say a number every time you hear a "beep" numbers between 1 to 10, randomly, you should avoid saying predefined sequences, for example 3, 4, 5, 6 or 9, 8, 7 or 3, 6, 9. Imagine a hat that contains 10 balls numbered from 1 to 10. When one of the balls is drawn, the number drawn is told and then the ball is returned to the hat for a new draw. *If the participant presents difficulty in understanding the test, an instruction adapted from that of young children will be given depending on the age/schooling of the participant. A practice run with a few numbers will be done to make sure that the participant has understood the test.*

□ **Updating words**

□ **Item 1**

Instructions: You are going to listen to some lists of words. These lists may have 4, 6, 8 or even 9 words. At the end of the list you have to **repeat the last 3 words you** heard.

For example if the list is "cat, balloon, dice and net", you should say: "cat"; then "cat, balloon"; then "cat, balloon, dice"; and finally "balloon, dice and net". At the end, only the last 3 words are left, the first and oldest, which was "cat", was eliminated, so that at the end of the task only the last 3 were left. Did you understand?

Important: Remember to repeat in the order you hear the words and always the last 3 words remain.

Shall we have a training session?

Itens de treino

SACO _ PIA _ AULA _ LAÇO

JOGO _ TREM _ TERRA _ MAR _ LUA

BEM _ SOLA _ MEL _ GEMA _ AVE _ BAÚ

□ ***Keep track***

□ **Item 1**

Instructions: In this activity you will see some pictures and they are divided into different groups. You will first see the name of the group and then the images that belong to this group.

□ **Item 2**

Instruction: Now you see the same images again and some new images and you will need to remember the last image from the categories I say. For example. (*do the training*). Do you understand? (*The instruction is the same for all lists*).

□ **Park Task**

Version 1: Imagine that you are going to visit a park, which is represented in this picture. In this visit you have to pass **6** specific **places in** the **shortest time possible**. You can go to other places different from these 6 that I am going to show you, but that will waste your time. This card with the instructions I am going to give you now will be available to you during the task.

You need to pass, in the order you wish, the **duck** pond (point at *map and instruction card*), the **popcorn** cart (point at map *and instruction card*), the **ice cream man (point at map** *and instruction card), the* **bike** area (point at map *and instruction card),* the **slide** (point at map and instruction card), and the **skateboard** area (*point at map and instruction card*).

You should **start at the gate** (*point on map and instruction card*) and **end at the toilet** (*point* *on map and instruction card*). The paths that have this **grid** (point on map *and instruction card*) are **free**, you can pass as many times as you want. **Blank paths** (point on map *and instruction card*) you can **only pass once**, you can't go back and forth on the same path if it's white. The **flower path** (point on map *and instruction card*), you can pass **only once**, if you pass only one part of it (*point on map and instruction card*), you can**'t** pass another part afterwards.

Remember to do the park tour, passing the 6 places, as quickly as you can. Do you have any questions? Can we get started?

Version 2: Now you will do the park visit **again**, but this time you will just follow this **order shown** (*point on the instruction card - **speak each item***). Remember that you need to follow **this order that is on the card**, which will be available to you during the test.

The rules are the same as before. You must **start at the gate** (*point on map and instruction card*) and **end at the toilet** (*point on map and instruction card*). The paths that have this **grid** (point on map *and instruction card)* are **free**, you can pass as many times as you want. **Blank paths** (point on map *and instruction card*), you can **only pass once**, but you can't go back and forth on the same path if it's white. The **flower path** (point on *map and instruction card*), you can pass **only once,** if you pass only one part of it (*point on map and instruction card*), you can**'t** pass another part afterwards.

Remember to visit the park by passing the 6 places, in the order indicated, as quickly as you can. Do you have any questions? Can we start?

Throughout the battery, at the end of each instruction given, the participant is asked to explain what they should do in the task, so there is assurance that they have understood the instruction and what will be needed to carry out the activity.

4.2 Stage II: Pilot Study 1 - Analysis of the Adequacy of the First Draft of the Battery

4.2.1 Methods

After the selection of the tasks and the construction/adaptation of their instructions, ten 6-year-old children from public schools (municipal or state) in the city of São Paulo were selected in order to verify if the instructions were understandable.

To explain the research objectives, parents were invited to a meeting; those who agreed with the participation of their child in the research were asked to sign the Term of

Consent (Appendix B), as well as to complete the health questionnaire (Appendix A). The children authorized by their parents/guardians were invited to participate in the study and also signed the Term of Consent (Appendix C).

Those who, as reported by the parents, were diagnosed with developmental disorders (intellectual disability, epilepsy, learning disorders, attention deficit/hyperactivity disorder) and/or uncorrected sensory deficits (auditory or visual), as well as delayed neuropsychomotor development were excluded.

The analyses were qualitative, carried out by the group of researchers, through extensive discussions about the initial collections, which involved the spontaneous feedback from each participant, participant observation during the performance of the tasks and the final responses of each subtest.

4.2.2 Results

According to the observations made during pilot study I, some changes were made in the instructions, only to facilitate the participant's understanding of the task. Below is a description of the changes, in relation to the qualitative analysis of the pilot study.

In the **category alternation** task, the following passages have changed:

- 1st block ○ "describes something living or non-living" - changed to: "is of something living or dead (lifeless)",
  - "(...) State as quickly as you can whether the image is of something living or non-living/dead..." - changed to: "(...) Say, as quickly as you can, whether the image is of something living or dead (lifeless)..."
  - "You will see the same pictures, but now you must say, as quickly as you can, whether the picture represents something big or small" - the word "represents" has been replaced by "is of";
  - "To know if the object is big or small, think about the size of each image compared to a football" - the expression "compared to" has been replaced by "close to"; ○ "(...) between saying whether the figure is something 'living or non-living/dead' or whether it is big or small..." - replaced by "alive or dead (lifeless)".
- 2nd block ○ **Excerpt without change**: "You will see the same pictures, but now you must say as quickly as you can whether the picture represents something big or small. To find out whether the object is big or small, think about the size of each picture

compared to a football: if it's bigger you can say 'big', if it's smaller you can say 'small'. You will repeat this task a few times.

Any questions? Can we begin?" ○ **Excerpt changed**: "You will see the same pictures but now you must say as quickly as you can whether the picture is of something big or small. To find out whether the object is big or small, think about the size of each picture near a football: if it is bigger you can say 'big', if it is smaller you can say 'small'. You will repeat this task a few times. Any doubts?

Can we start?" ○ "between saying whether the figure is something living or "not living/dead" - changed to: "dead (lifeless)". That is, in the first figure, if you say

whether it is 'alive or not alive' has been changed to 'alive or dead'; in the next picture, you must say whether it is 'big or small', and so on...

- 3rd block ○ No **change**: "You will see the same pictures that you saw before, but this time you must alternate between the classifications, between saying if the picture is something alive or not alive (when it is not alive you can say - dead) or if it is big or small. That is, in the first picture if you say whether it is 'living or non-living', in the next picture you must say whether it is 'big or small', and so on. You will have to remember which classification you used in the previous picture to be able to classify the next picture. You will repeat this task a few times. Any questions? Can we get started?"
  - **Excerpt changed**: "You will see the same pictures that you saw before, but this time you must alternate between the classifications, between saying if the picture is something living or dead (lifeless) or if it is big or small. That is, in the first picture if you say whether it is 'alive or dead (lifeless), in the next picture you must say whether it is 'big or small', and so on. You will have to remember which classification you used in the previous picture to be able to classify the next picture. You will repeat this task a few times. Any questions? Can we get started?"

The change from "not alive" to "dead" was made because of the answers given by the participants themselves, who, when instructed to say "not alive", automatically said "dead"; with this, a conflict was created in the response, sometimes they said not alive, sometimes they said dead and also corrected themselves after saying dead for not alive, as instructed, which caused more time in performing the task. This change then aimed to facilitate the understanding of the task, removed a possible conflict and also reduced the total time of completion, by saying dead / alive instead of alive / not alive.

As for the **go/no-go** task, a new sentence was inserted: "But it is important that you say yes to the other numbers, if you don't it will be considered an error."

- **Unchanged**: "In this task you will hear some numbers and I need you to say YES to every number you hear, EXCEPT number 4. When you hear 4, don't say anything, be quiet. Got it? So let's have a practice run."
- **Excerpt changed**: "In this task you will hear some numbers and I need you to say YES to every number you hear EXCEPT (minus) number 4. When you hear 4, don't say anything, be quiet. But it is important that you say yes to the other numbers, if you don't, it will be considered a mistake. Do you understand? Then let's have some practice."

In the ***keep track*** task, a sentence was also inserted: "Remember: at the end of the list say the last picture in each group, the last toy you saw, the last kitchen item and the last outfit or accessory you can remember."

- **Excerpt without change**: "In this activity, you will see some images and they are divided into different groups. First you will see the name of the group and then the images that belong to this group.
  Now that you know the images of the 3 categories (TOY, KITCHEN and CLOTHING/ACESSORIES), you will need to remember the last figure of each of these groups. In addition to these figures that you have seen now, new ones will appear."
- **Excerpt changed**: "In this activity, you will see some images and they are divided into different groups. First you will see the name of the group and then the images that belong to this group.
  Now that you know the pictures from the 3 categories (TOY, KITCHEN and CLOTHING ACCESSORIES), you will need to remember the last picture from each of these groups. In addition to the pictures you have seen now, new ones will appear. Remember: at the end of the list say the last figure in each group, the last toy you saw, the last kitchen item and the last clothes or accessory you can remember."

4.3 Stage III: Pilot Study 2 - Analysis of the Adequacy of the Second Version of the Battery

4.3.1 - Methods

After the analysis of pilot study 1, which generated the adjustments in the instructions, another 12 children were selected, ten with 6 years old and two with 12 years old, all enrolled in public schools in the city of São Paulo. This pilot study aimed to analyze possible ceiling and floor effects.

In the same way, as done in the pilot study 1, parents were invited to a meeting to explain the objectives of the research; those who agreed with the participation of their children in the research were asked to sign the Term of Consent (Appendix B), as well as to complete the health questionnaire (Appendix A). Children authorized by their parents/guardians were invited to participate in the study and also signed the Consent Form (Appendix C).

Those who, according to the parents' report, have been diagnosed with developmental disorders (intellectual disability, epilepsy, learning disorders, attention deficit/hyperactivity disorder) and/or uncorrected sensory deficits (auditory or visual), as well as delayed neuropsychomotor development were excluded.

4.3.2 Results

The analyses, qualitative and quantitative, dealt with the difficulty of execution and the possible effects of floor or ceiling on performance and were carried out by the group of researchers, through extensive discussions about the initial collections that involved the process of performing the tasks and the final responses of each participant in the subtests.

It was found that some of the tests described above should undergo changes, be excluded or replaced, as they were not suitable for the purpose of assessing the FE domain or presented ceiling effect on performance.

□ ***Colour-shape***

In this task, the images were kept and in the same order; however, before, in block 1, the "Tic Tac Toe" indicated that the participant should say the shape of the figure; and, in block 2, the rainbow indicated that the participant should say the colour of the figure. Thus, only one small change was made: the track symbols were removed to increase the cognitive demand in the evaluation of the specific domain (Figure 18).

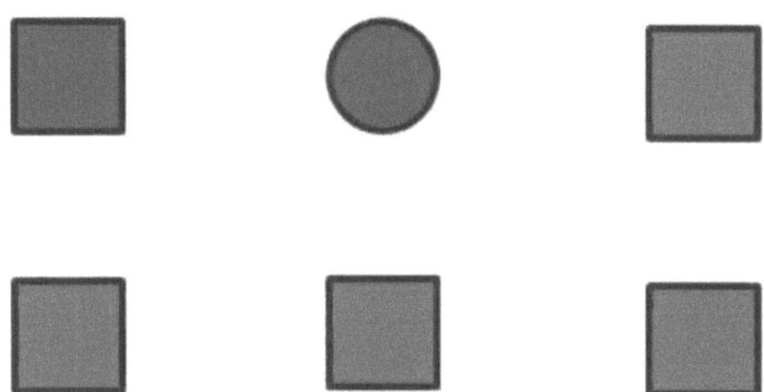

Figure 18- Change in the colour-form task
Source: Prepared by the researchers of the broad project.

□ **Category *switch***

The first change in this task was made in the same way as in the above mentioned task: the track-symbols (arrows and heart) were excluded. However, after pilot study 2, other changes were necessary.

Some tests were excluded due to their inefficiency, in the evaluation of the specific domain, according to the data collected or due to the difficulty of understanding and performing the task by the participants. The changes and adjustments were as follows:

a) Images: it was found that some were not part of the repertoire of the population to be studied as frequently as others available. They were replaced as follows: dinosaur for elephant; bridge for circus; preacher for whistle; and snail for worm;

b) Clue symbols: previously, there was either a cross with arrows, indicating the category large or small, or a heart to indicate the category living or non-living. These symbols were removed, due to the ease of execution and to differentiate it from the competing task.

c) Presentation: the images that were previously presented one at a time, in the centre of an A4 sheet, were placed in a single sheet, also of A4 size, four lines with five images each, in phases 1 and 2 of the test (large/small or dead/live); the third phase (alternation) was composed of eight lines with five stimuli in each line.

Figure 19 - Dead/live phase
Source: Prepared by the researchers of the broad project.

Figure 20 _ Large/small phase
Source: Prepared by the researchers of the broad project.

Figura 22 _ Fase alternância (parte 1)

Figura 21 _ Fase alternância (parte 2)

Figure 22 _ Phase alternation (part 1)
Figure 21 - Phase alternation (part 2)
Source: Prepared by the researchers of the large project

□ ***Go/No-go***

Before being altered, this task was composed of three blocks, from which the two initial blocks were excluded. It was found that only part of the task was necessary to assess the measured domain; therefore, only half of the third block was kept, enough to assess the response inhibition function. It also had execution time as a measure, which was also excluded, since it was not possible to measure this variable, since the task itself establishes execution time.

1	5	4	8	3	9	4	2	5	7
0	4	3	1	5	7	2	6	4	8
3	5	7	2	4	3	1	5	9	4
5	7	6	4	3	1	4	2	6	0
8	4	6	1	9	3	0	4	2	5
0	7	4	9	8	4	1	6	3	7
4	0	7	8	0	6	4	5	2	1
6	8	3	4	5	9	7	2	4	0
2	6	8	0	3	4	5	7	1	4
9	1	4	6	5	0	2	4	3	8

Figure 23 - *Go/No-go* task
Source: Prepared by the researchers of the broad project.

□ ***The stroop-numerical test***

Task excluded from battery as there was performance ceiling effect in the way we elaborated, for all ages.

□ **Updating words**

This task has undergone many changes and adaptations. It was composed in study 1 only of monosyllable figures, also changing to disyllable ones due to the difficulty of finding the necessary number of images, without them being repeated.

Afterwards, the presentation was changed from visual to auditory task, with monosyllable and disyllable words. The presentation time for each image, which was 5 seconds, was changed to a 2 second interval between each word and the number of stimuli to be remembered after the update went from three to four. At the end of each word list presented, a card with four spaces is shown for the participant to indicate the order in which he/she remembers the words.

□ **Park Task**

The alteration of this task was in the *Free path* (where you can go and return with the pencil as many times as you want) and *Only 1 time* (where you are allowed to pass only 1 time), in the matter of colour. The *Free* path was gray and the *Only 1 time* path, white, colours that were inverted to facilitate the compression of the task and reduce cognitive demand.

This was necessary because, for the children, it was confusing to see that the blank path was not allowed to pass several times, while the grey path was. It was decided to leave the *Free path* without any colour stimulus, so that, when looking at it, the child would know that they could pass through it as many times as they thought necessary, and the *Only 1 time* path, with the grey stimulus, indicating that a second passage would be forbidden.

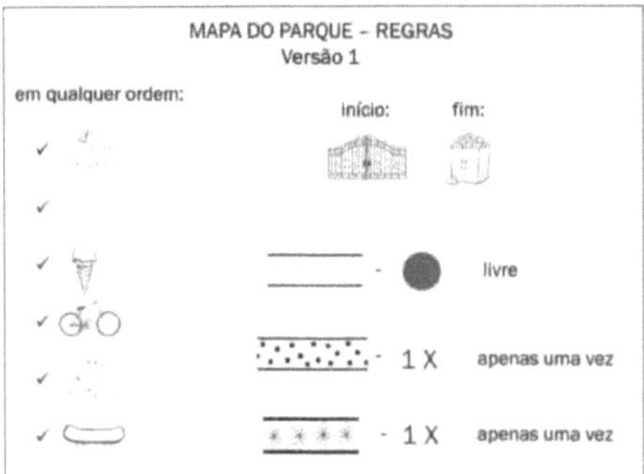

Figure 24 - Park task (1ª rule)
Source: Prepared by the researchers of the large project

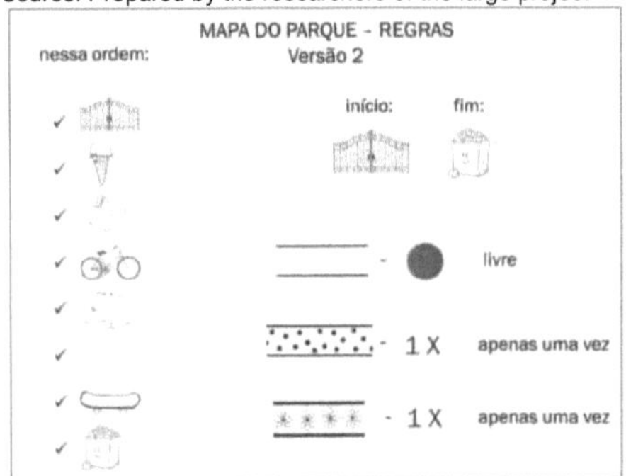

Figure 25 - Park task (2ª rule)
Source: Prepared by the researchers of the broad project.

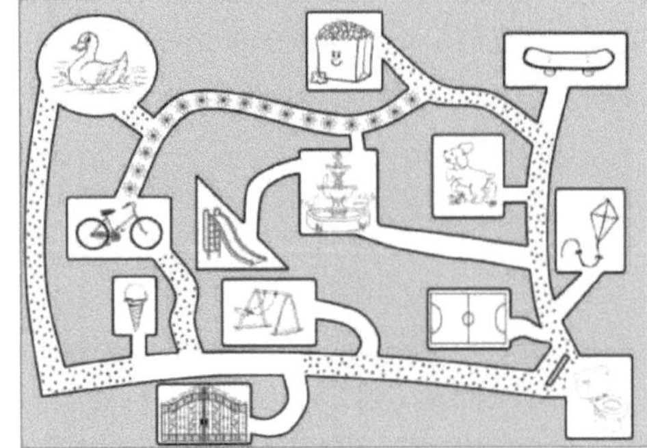

Figure 26 - Park map

Source: Prepared by the researchers of the broad project.

□ ***Dual task***

Adapted from Fournier-Vicenti (2008), this task was included in the present research without first being tested in the pilot study because, by the start date, its investigation had not been completed.

The dual-task paradigm consists of a paper-and-pencil test involving a visuospatial (cross circles) and phonological/verbal (*span* digits) grip task.

The digit task consists of, for 90 seconds, repeating sequences of orally presented digits with the number of digits corresponding to each individual's digit length. To determine participants' digit span, lists of digits are read aloud at the rate of one digit per second, and participants are asked to repeat them in the order of presentation, the number of digits increasing progressively. Participants' *span* is the maximum length at which they correctly repeated five out of six digit sequences.

The circles task (Figure 27) consists of traversing a chain of 240 circles connected with arrows to form a path drawn on a sheet of paper, A4 size. Subjects are required to follow the path as quickly as possible over a period of 90 seconds. There is a training item for this task before the test begins.

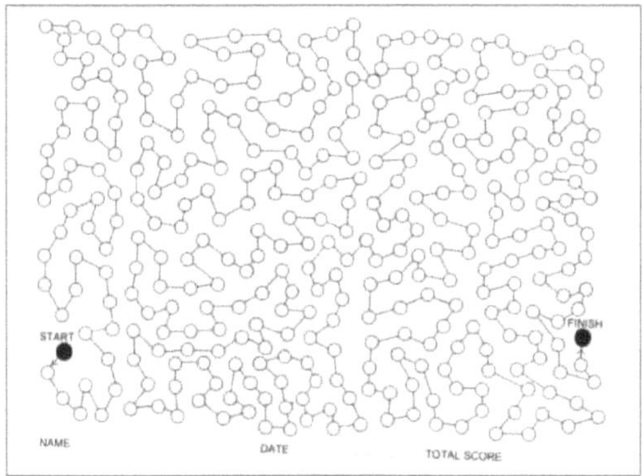

Figure 27 - Circles (paths) task
Source: Baddeley et. al. (1997)

The dual task condition consists in the simultaneous execution of the two tasks within 90 seconds. To quantify the participant's performance, the mu index is used (BADDELEY et al., 1997), which expresses the global percentage of success in the dual task in relation to the individual tasks.

***Dual* task instructions**

- **Determination of *span***: I am going to say some numbers and you must repeat as soon as I finish speaking. Try to repeat in the same order, as fast as you can.
- **Memory to list (*single task*):** Again you must repeat the numbers as soon as I finish speaking. Try to repeat in the same order, as fast as you can.
- the **Path of circles (*single task*):** Now you will move from one little ball to the other by following the lines that are between them. Here I will draw a "wall" (between the balls together), just to remind you that you cannot go through here, because there is no line (*show the right path*). Do this as quickly as you can. This is a practice drawing.
- Now it's for real. Do the same thing you did before, as fast as you can. Stop only when I say so.

- **Dual task:** This time, we're going to do both activities at the same time. While I'm speaking the numbers and you repeat, as we did at the beginning of this activity, you should also trace, make the path of the little balls, this same activity you just did. Both at the same time. Got it? Can we start?

The RNG and Verbal Fluency tests have not changed since the beginning of the research.

Thus, at the end of the pilot studies, the selection of tests of the FE battery was composed by the following assessment instruments (Chart 2):

Chart 2 - Tests that compose the Fes battery

Specific domain of FEs	Task	Scores
Alternation	Colorform	Task composed of three phases. At the end it is verified the cost of errors and time in completion.
	Alternation of categories	Task composed of three phases. At the end it is verified the cost of errors and time in completion.
Inhibition	*Go/No-go*	*Errors of action, errors of omission.*
Update	*Keep track*	*At the end of the 8 lists, the maximum score can reach 28.*
	Updating words	After 12 sequences, *the maximum score can reach*

		48 words.
	RNG	Randomness indices are calculated.
Planning	Park Task	Performance is calculated by the cost of errors and time in performing the first and second phase of the task.
Dual task	*Dual task*	*Performance on this task is quantified using the UM index (Baddeley et al.,1997)*

Source: Elaborated by the author

At the end of the construction of the FE battery, the research group decided to test the battery in a clinical group of children with ADHD, a disorder that may be associated with motor, perceptual, cognitive and behavioural disorders, expressing global difficulties in child development (RIZZUTTI et al., 2008).

These symptoms and characteristics can be explained by alterations in the FEs (COZZA, 2005, ASSEF, 2005, ROHDE; MATTOS, 2003). Thus, children with this disorder tend to present great difficulties in various contexts and aspects of their lives, especially at school.

Because it compromises FEs and other cognitive functions, the diagnosis of ADHD in children is not an easy task, so new neuropsychological assessment procedures may favor diagnostic accuracy. In this sense, for assessment in childhood, it is important to have instruments that assess, in addition to isolated skills, the various cognitive functions, with the possibility of observing the interrelationship between the different neuropsychological functions (ARGOLO et al., 2009).

4.4 Step IV: Clinical Sample Study - Children with ADHD

4.4.1 Casuistry

The sample was recruited from the Outpatient Clinic of the Núcleo de Atendimento Neuropsicológico Infantil Interdisciplinar (NANI), which uses a specific protocol for diagnosing children with ADHD, consisting of an interdisciplinary assessment - neurological, psychiatric and neuropsychological (RIZZUTTI et al., 2008)

The diagnosis is established from medical and neuropsychological assessment using the following instruments:

- Intellectual level: estimated IQ, using the Wechsler Intelligence Scale for Children (WISC-III);
- Attention assessment: Conners' Continuous Performance Test (CCPT); Symbol Cancellation and Auditory Attention, assessed by means of the NEPSY-II battery (ARGOLO, 2010);
- Working Memory: assessed by means of the AWMA battery, reduced, consisting of the subtests - Digit Recall, Listening Recall, Block Recall, Spatial Recall, Counting Span (ALLOWAY, 2007). Corsi Blocks (direct and reverse recall) and List of Words - Interference: battery
  NEPSY II (ARGOLLO 2010);
- Episodic memory and visuo-constructive function: copying and recalling Rey's Complex Figure;
- Executive Functions: Drawing Fluency, Inhibiting Responses - NEPSY II battery (ARGOLLO, 2010);
- Behavior Rating Inventory of Executive Functions (BRIEF): questionnaire answered by parents or teachers about the frequency of certain behaviors associated with FE (behavior regulation, metacognition and global FE) (Carim et al., 2012)
- Social perception - Theory of Mind (contextual task), assessed by means of the NEPSY II battery (ARGOLO, 2010);
- Behaviour: Child Behaviour Checklist (CBCL); EACI-P scale answered by the parents;
- Neurological and psychiatric evaluation: clinical evaluation, DSM-IV diagnostic criteria, applied by means of a questionnaire answered by parents/guardians; detailed anamnesis.

Hyperactivity conditions secondary to intellectual disability, epilepsy and/or the presence of uncorrected sensory deficits (auditory or visual) were excluded.

The Control Group, consisting of 17 participants, recruited from public and public schools, was paired with the ADHD Group by age, gender and type of school.

Children and adolescents with an age compatible with the current school year, with no history of repetition and authorized by parents/guardians by signing the Informed Consent Form participated in the study. Also excluded were those who, according to their parents' reports, presented diagnoses of developmental disorders (intellectual disability, epilepsy, learning disorders, attention deficit/hyperactivity disorder) and/or uncorrected sensory deficits (auditory or visual), as well as delayed neuropsychomotor development.

For this, the following procedures were adopted: a) a questionnaire was sent to the participants' parents/guardians, which covers the investigation of socioeconomic data, health aspects and schooling history (Appendix A); b) the abbreviated Conners scale (BRITO, 1987) was applied, which, comprising ten items, assesses behavior problems such as hyperactivity, inattention, among others - such scale was standardized for the Brazilian population and has been widely used as a screening tool in the selection for research on ADHD patients -; children who presented a score above the cut-off point for their age were excluded from the sample.

4.4.2 Evaluation of Executive Functions

The instruments used in this phase aim to analyze the cognitive performance and FEs in children with ADHD. Thus, from the broad evaluation performed for diagnosis, already described, the following test results were selected:

- *Counting Recall*, from the AWMA battery, which assesses verbal operational memory. To perform this task, the child must count the number of circles, where triangles are also displayed on a computer screen. Then, they must recall, in the correct order, the number of circles in each group. The test starts with only one group and can reach the series with seven groups.
- Estimated WISC-III IQ (vocabulary and cubes).

- Information subtest: additionally applied, it is part of the verbal scale of the WISC-III battery and assesses the extent of knowledge acquired, the quality of formal education and motivation for school achievement. It was included in the research in order to verify whether it is possible to correlate the stimulation of the school environment with FE.
- Verbal fluency task (phonological and semantic): has been widely used in the evaluation of FEs. The responses depend on the level of intelligence, vocabulary and attention, and operational memory components are also necessary for the individual not to persevere in the responses (BUTMAN et al., 2000).

At this stage of the study it was decided to use only one task per domain, since for the ADHD group it would have a significant fatigue effect. The tasks that comprised the FE battery in this phase are shown in Chart 3.

Table 3 - Tasks for assessing FEs, separated by specific domains

Specific domain of FEs	Task
Access to long-term memory	Verbal fluency
Operating memory	*Counting recall*
Alternation	Alternation of categories
Inhibition	*Go/No-go* | RNG
Update	Updating words | RNG
Planning	Park Task
Dual task (central executive - verbal/visuospatial)	*Dual task*

Source: Elaborated by the author

4.4.3 General procedures

The evaluations of both groups were individual: the ADHD Group was evaluated at Nani, in a silent room specific for this purpose; the applications of the Control Group were carried out at the school institutions attended by the children and adolescents; and the evaluation of the children recruited by indication was carried out at a place convenient to parents and children (their own homes), but in an adequate environment.

With the authorization of the schools, the children of each school year were selected; then, the parents were called for a meeting, in order to explain the research objectives, and those who agreed with the participation of their children were asked to sign the Term of Consent (Appendix B), as well as to fill out the health questionnaire (Appendix A). The children authorized by their parents/guardians were invited to participate in the study and signed the Term of Consent (Appendix C).

The battery of tests was applied individually at the school, in a room adequate for sound and light, in a single session. The subtests were applied in a randomized order in order to avoid factors that could influence the results.

4.4.4 Statistical Analysis

Data were analysed using the Statistical Package for Social Sciences (SPSS) for Windows - with a significance level of $p < 0.05$.

Initially, descriptive analyses were performed to assess frequency, mean, and standard deviation of the variables of interest. In order to test the difference in means in the ADHD and Control Groups, we used the T-Student Test for independent samples, verifying the following assumptions: homogeneity of variance and independence of measures. Effect magnitude indices were also calculated (*effect sizes; Cohen d), in* order to demonstrate the absence or not of effects in certain tests due to the small sample size. The effect size is a descriptive statistic that serves as a complement to the test of statistical significance.

4.4.5 Results and Discussion

The total sample of this study was composed of 34 children, of which 17 were part of the ADHD Group and the other 17 were part of the Control Group, both paired with respect to gender, age and type of school.

In this sample, 76.5% (26) of the individuals were male; age ranged from 6 to 12 years, with a mean of 9 years (SD = 2.3); regarding the type of school, 70.6% (24) of the children attended public schools (Table 1).

Table 1 - Sample distribution according to the pairing in relation to gender and type of school

		Quantity	%	Grand total (%)
Group	ADHD	17	50	100
	Control	17	50	
School	Private	10	29,4	100
	Public	24	70,6	
Gender	Male	26	76,5	100
	Female	8	23,5	

Source: Elaborated by the author

Table 2 shows the results of the IQ evaluation, information subtest and FE battery. There was no statistically significant difference between the groups regarding IQ (p = 0.19) and also in the information subtest (p = 0.15).

Table 2 - Comparison of means and standard deviation of FEs and Complementary tasks between the ADHD and Control Groups

Tasks	ADHD		CONTROL				
	Average	DP	Average	DP	t (df)	p	Cohen d
Estimated IQ	103,3	27,00	112,76	12,09	0,22 (31)	0,19	0,45
Information (weighted)	9,06	3,41	10,53	2,45	0,26 (32)	0,15	0,49
Category switching (cost of errors)	10,41	8,48	5,18	5,18	2,17 (32)	0,03	0,74
Category change (time cost)	68,47	43,62	61,65	31,09	0,52 (32)	0,60	0,74
Word update (random)	25,71	5,88	25,18	6,60	0,72 (32)	0,80	0,08
Updating words (sequence)	10,88	5,89	14,35	7,06	0,74 (32)	0,13	0,53

***Counting* (memory score)**	88,63	10,03	94,16	22,97	-0,9 (32)	0,37	0,31
***Counting* (processing score)**	90,96	9,75	97,89	15,00	-1,59 (32)	0,12	0,54
Dual task	105,2	15,00	94,14	14,7	2,17 (32)	0,03	0,74
***Go/No-go* (errors of**	3,35	3,352 ,47	1,46	1 (23)	0,29	0.36 **share)**	

Table 2 - Comparison of means and standard deviation of FEs and Complementary tasks between the ADHD and Control Groups

	ADHD		**CONTROL**				
Tasks	**Average**	**DP**	**Average**	**DP**	**t (df)**	**p**	**Cohen d**
***Go/No-go* (errors of omission)**	2,94	5,59	1,35	2,23	1 (20)	0,28	0,37
RNG Evans	0,33	0,10	0,33	0,08	0,49 (32)	0,81	0,08
Park Task (error cost)	3,06	4,50	2,18	2,27	0,65 (32)	0,47	0,24
Park Task (time cost)	65,47	53,60	53,82	45,71	0,07 (32)	0,50	0,23
Phonological fluency (S)	5,88	2,39	5,65	2,23	0,29 (32)	0,76	0,66
Phonological fluency (F)	7,53	2,98	5,71	2,51	1,92 (32)	0,63	0,09
Semantic fluency (animals)	14,12	4,31	11,24	3,12	2,21 (32)	0,34	0,76
Semantic fluency (fruits)	9,71	3,91	8,94	2,33	0,69 (32)	0,49	0,23

Regarding the results of the FEs battery, in the alternation domain (task alternating categories), the comparison of scores (alternation cost) between the ADHD and Control Groups showed statistically significant differences only in the cost of errors (p = 0.03; d = 0.74). The ADHD Group showed the worst performance, with no significant differences in the cost of time (p = 0.6; d = 0.74).

In the updating domain (word updating task), in the randomness score, the mean values for both groups did not show statistically significant values (p = 0.8; d = 0.08) and neither did the sequence score (p = 0.13; d = 0.53).

In the working memory domain (*counting span*) there were also no statistical differences in memory (p = 0.37; d = 0.31) and processing (p = 0.12; d = 0.54) scores.

In the dual *task* domain, there was proven statistical significance (p = 0.03; d = 0.74), in which the ADHD Group presented a better performance.

In the inhibition domain (*go/no-go*), the mean values for both groups showed no statistical significance, both for action errors (p = 0.28; d = 0.36) and omission errors (p = 0.29; d = 0.37).

In the random number generation task (inhibition and updating), the groups showed similar means, with results without statistical significance (p = 0.81; d = 0.08).

In the planning domain (park task), in both calculated scores (time and error cost), the groups did not show statistically significant values (time cost: p = 0.50; d = 0.23 and error cost: p = 0.47; d = 0.24).

In the verbal fluency domain (access to long-term memory), a task involving four phases - two of phonological fluency (F and S) and two of semantic fluency (animals and fruits) - each item was calculated separately. The means were not statistically significant: F: $p = 0.63$; $d = 0.66$; S: $p = 0.76$; $d = 0.09$; and semantic fluency - animals: $p = 0.34$; $d = 0.76$; fruits: $p = 0.49$; $d = 0.23$.

In general, in this study, few tasks showed results that differentiated the group of children with ADHD from children with typical development. Most of them, such as the tasks category alternation (time cost), word updating, *counting span, go/no-go,* random number generation, park task and verbal fluency, did not differentiate the groups; only the tasks alternation (category alternation task) and dual task differentiated the groups.

This may be due to the small sample size, consisting of only 17 children in each group, besides the fact that the battery is still being tested. Or even, to the fact that the battery is not yet sensitive to differentiate the groups.

CHAPTER 5

GENERAL DISCUSSION

The main purpose of this three-stage study was to develop and propose a battery that assesses six distinct domains in FEs, so that it can be used with various participants of different age groups, from childhood to adulthood, and also by different intellectual levels, prioritising the assessment of children and adolescents (6 to 12 years old).

There is evidence of the relationship between FEs and school learning - reading, mathematical skills and concepts - (LIMA; SALGADO; CIASCA, 2008, CLARK; WOODARD, 2010). Thus, new instruments and procedures to assess FEs may help in the knowledge and understanding of specific domains and their deficits, as well as assist the diagnosis of various disorders.

Given the difficulties already mentioned to accurately assess FEs (MIYAKE et al., 2000), that is, some instruments currently used end up assessing more than one domain at the same time and thus it is not possible to identify which function or domain is preserved or compromised.

In light of this, this paper documents the existence of six domains of FEs, using as a referential basis Miyake et al. (2000), who suggest that the three frequently postulated FEs (alternating, updating, inhibition) are separable but moderately correlated constructs, thus indicating the unity and diversity of FEs. The other three domains were based on: Fisk and Sharp (2004) - access to long-term memory; Owen (1997) - planning; and Shallice (1982) - dual tasking.

To assess these domains, the battery was organized in seven trials, being two for the alternation domain (colour-form and category alternation); two for updating (*keep track* and word updating); and one for each of the other domains: inhibition (*go/no-go*), planning (park map); and dual task (dual task).

For each function, it was possible to obtain the score of each test independently, without the need to administer the entire battery. When adapting the battery, we tried to meet the following criteria: accessible and understandable language for different socioeconomic strata and school years; non-inclusion of written material or that could require reading ability; high frequency verbal stimuli in Portuguese language (PINHEIRO, 1996); non repetition of stimulus or category of stimuli in different tests; free use material -

not commercial.

The tests were selected and adapted from other studies carried out with children, adolescents and adults. For the selection of each subtest, a qualitative and quantitative analysis was performed by the group of researchers, in order to verify facilities and/or difficulties of understanding in the instructions and execution of the tasks, and possible effects of floor or ceiling (ceilling effect/ floor effect) in the performance.

The **color-form** task was adapted from Miyake et al. (2004) and Friedman et al. (2008); and the **category alternation**, from Mayr and Kliegl (2000), Friedman et al. (2008). Several researchers have used these tests (FRIEDMAN et al., 2006, FRIEDMAN et al., 2007, 2008) and have successfully demonstrated the unity and diversity of FEs. The color-form task will be tested still in this format. The category alternation task and the comparison of scores (alternation cost) between the ADHD and Control Groups showed statistically significant differences only in the cost of errors, in which the Control Group showed better performance. This may indicate task adequacy.

Regarding the tasks of the **update** domain (*keep track,* adapted from Miyake et al., 2000, Yntema, 1967) and the word **update** domain (*letter* memory), adapted from Morris and Jones (1990), some authors carried out studies with adults (MIYAKE, 2000, FRIEDMAN et al. 2006, FRIEDMAN, 2007, 2008) and others with children and adolescents (ST CLAIR-THOMPSON GATHERCOLE, 2006, TAMNES et al., 2010). These are tasks that require monitoring and coding the information received, followed by an appropriate review of the items kept in the working memory, replacing old and irrelevant information with more recent and relevant information. It is emphasized that this domain goes beyond the simple maintenance of information relevant to the performance of the task, because it requires a dynamic manipulation of the contents of the working memory (LEHTO, 1996, MORRIS; JONES, 1990). That is, updating is centred on the need to actively manipulate relevant information in working memory, rather than just passively storing it. Both should also be analysed regarding the current format.

The **dual task** - adapted from Fournier-Vicenti (2008) and Della Sala et al. (2010), was selected because it is a simple pencil-and-paper test, easy to use and cheaper than existing computerized methodologies. Also because, according to Della Sala (2010), the test-retest reliability is sufficiently high making it ideal for clinical use. This task also eliminates the possibility of measurement errors, avoiding the involvement of other cognitive processes, which is the proposal of this work, besides minimizing the effect of any cultural or educational variation. Della Sala (2010) did not identify any age effect, suggesting that the cognitive function involved in the simultaneous performance of two distinct tasks is not affected by healthy aging (BADDELEY et al., 1986). The dual task brought statistically significant results, this time with the ADHD group showing better performance.

The ***go/no-go*** test, adapted from Casey et al. (1997) was chosen basically for the same reasons as the previously mentioned task, that is, because it is a frequently used task whose simple format allows the analysis of response inhibition under conditions in which other cognitive/behavioral processes are minimized. This task should also be reviewed, regarding the number of stimuli and the interval between stimuli.

It is believed that the **random number generation** task, adapted from Badelley (1966), is a useful measure for clinical and experimental practice to investigate FE. However, further studies are needed because, according to Miyake et al. (2000) - the main theoretical source of this paper - this task seems to depend on multiple skills, especially the inhibition and updating abilities, which ends up not respecting the aim of this research, which proposes pure measures for the assessment of specific domains. However, at a broad level, our research around this task continues, with different age groups, school levels and pathological conditions, so that the objective can be achieved later. In this study, we chose to use it in the evaluation of the **updating** domain, that is, in monitoring the distribution of responses (numbers).

In the **planning** domain, we included the **park task** - adapted from Wilson et al. (1996), from the BADS Battery, originally for use with people aged 7 years and older. In it, participants must visit a series of designated locations on a map of a park; however, when planning the route, rules must be followed. The comparison of the two test results allows us to assess children's planning ability and its ecological validity is an advantage for those who wish to examine everyday difficulties in executive functioning. Another issue lies in the fact that it is a non-verbal task, which may be an advantage when testing certain populations with these difficulties, of reading and writing (ENSLIE et al., 2003). It is believed that the correction criteria regarding time should be reviewed.

All tasks were adapted according to the demand and need of each participant. Words, figures, and images were removed and replaced, after careful analysis and discussion of the researchers, respecting the purpose of evaluation of each subtest.

Thus, this battery is still under development, new changes will be necessary and others are already being carried out by researchers involved in this project, with children, adolescents and adults.

CHAPTER 6

CONCLUSIONS

Taking into account the relationship between FEs and learning in children, it is essential to understand how difficulties in FEs relate to these individuals. Despite the progress observed in relation to the assessment of FEs, there are still few instruments developed especially for the assessment of specific domains.

The finding that FEs show both unity and diversity has important theoretical and methodological implications for neuropsychological research. This study presents characteristics considered necessary for evaluation: its proposal is a battery of tests - and not a single test - that can be used by different individuals, of various age groups and intellectual levels. In addition, it is divested of any computerized equipment, which facilitates its use even in places that do not have such equipment and, finally, it is intended to be of free use, for all professionals.

Although there is no gold standard for assessing FEs, many tests are already available to assess the components of this complex construct. However, it is interesting to highlight the relevance of developing pure measures, thus, with the advancement of theories and FE models, tests can be more specific in the domains and components of FEs, making the assessment more accurate, sensitive and specific regarding each executive domain, helping in the diagnostic process.

REFERENCES

ALLAIN, P. et al. Executive functioning in normal aging: a study of action planning using the Zoo Map Test. **Brain Cogn**, v. 57, p. 4-7, 2005.

ANDERSON, P. Assessment and development of executive function (EF) during childhood. Neuropsychology, Development, and Cognition. Section C, Child Neuropsychology, v. 8, n. 2, p. 71-82, 2002.

ANDERSON, V. Assessing executive functions in children: Biological, psychological, and developmental considerations. **Developmental Neurorehabilitation**, v. 4, p. 119-136, 2001.

ARGIMON, I. I. L; BICCA, M.; TIMM, L. A.; VIVAN, A. Executive functions and the assessment of flexibility of thought in the elderly. **RBCEH - Revista Brasileira de Ciências do Envelhecimento Humano**, Passo Fundo, p. 35-42, jul./dez. 2006.

ARGOLLO, N. NEPSY II: Avaliação neuropsicológica do desenvolvimento. In: MALLOY-DINIZ, L. F.; FUENTES, D.; MATTOS, P.; ABREU, N. **Avaliação Neuropsicológica**.

Porto Alegre: Artmed, 2010.

BADDELEY, A. D. **Working memory, thought and action**. Oxford University Press: Oxford, 2007.

_____. **Working memory**. Oxford: Oxford University Press, 1986.

_____ HITCH, G. Working memory. In: Bower, G. H. (Ed.). **The psychology of learning and motivation**: Advances in research and theory. New York: Academic Press, 1974. p. 47-89. (v. 8)

_____. et al. Testing central executive functioning with a pencil-and-paper test. In: RABBITT, P. M. (Ed.). **Methodology of frontal and executive function**. Psychology Press: Hove, UK, 1997.

_____. Exploring the Central Executive. **The Quarterly Journal Of Experimental Psychology**, v. 49a, n. 1, p. 5-28, 1996.

_____; DELLA SALA, S. Working memory and executive control. **Philos Trans R Soc Lond**, v. 351, p. 1397-1403, 1996.

BARKLEY, R. A. **Attention deficit hyperactivity disorder** (2nd ed.). New York: Guilford Press, 1998.

_____. Behavioral inhibition, sustained attention and executive function: Constructing a unifying theory of ADHD. **Psychological Bulletin**, v. 121, p. 65-94, 1997.

BERK, L. E. **Child Development**. 7th ed. Boston: Pearson, 2006.

BEST, J. R.; MILLER, P. H. A Developmental Perspective on Executive Function. **Child Development**, University of Georgia, v. 81, n. 6, p. 1641-1660, Nov./Dec. 2010.

BRITO, G. N. O. The Conners abbreviated teacher rating scale: development of norms in Brazil. **Journal of Abnormal Child Psychology**, v. 15, n. 4, p. 511-518, 1987.

BURGES, P. W.; ALDERMAN, N. Executive dysfunction. In: GOLDSTEIN, L.; MCNEIL, J. (Org.). **Clinical Neurop- sychology**: A practical guide to assessment and management for clinicians. England: John Wiley & Sons Ltd, 2004. p. 185-270.

BUTMAN J.; ALLEGRI, R.F.; HARRIS, P.; DRAKE M. Fluencia verbal en Espanol. Datos normativos en Argentina. **Medicina**, n. 60 (5/1), p. 561-4, 2000.

CARIM, D. B., MIRANDA, M. C., Bueno, Orlando F A. Translation and adaptation to portuguese of the Behavior Rating Inventory of Executive Function - BRIEF. **Psicologia: Reflexão e Crítica** (UFRGS. Impresso). , v.25, p.653 - 661, 2012

CASEY, B. J. et al. A developmental functional MRI study of prefrontal activation during performance of a go-no-go task. **Journal of Cognitive Neuroscience**, v. 9, n. 6, p. 835-847, 1997.

CECI, S. J. How much does schooling influence general intelligence and its cognitive components? A reassessment of the evidence. **Developmental Psychology**, v. 27, n. 5, p. 703-722, 1991.

CHAN, R. C. et al. Assessment of executive functions: review of instruments and identification of critical issues. **Arch Clin Neuropsychol**, v. 23, p. 201-216, 2008.

_____. Assessment of executive functions: review of instruments and identification of critical issues. **Arch Clin Neuropsychol**, v. 23, p. 201-216, 2008.

CLARK, C. A. C.; PRITCHARD, V. E.; WOODWARD, L. J. Preschool executive functioning abilities predict early mathematics achievement. **Developmental Psychology**, v. 46, n. 5, p. 1176-1191, Sep. 2010.

COHEN, J. **Statistical power analysis for the behavioral sciences.** 2nd ed. New Jersey: Lawrence Earlbaum Associates, Hillsdale, 1988.

COLLETTE, F. et al. Exploration of the neural substrates of executive functioning by functional neuroimaging. **Neuroscience**, v. 139, p. 209-221, 2006.

_____. Involvement of both prefrontal and inferior parietal cortex in dual-task performance. **Brain Res Cogn Brain Res**, v. 24, p. 237-251, 2005.

CONKLIN, H. M.; LUCIANA, M.; HOOPER, C. J.; YARGER, R. S. Working memory performance in typically developing children and adolescents: Behavioral evidence of protracted frontal lobe development. **Developmental Neuropsychology**, v. 31, p. 103-128, 2007.

COZZA, H. F. P. **Avaliação das Funções Executivas em Crianças e Correlação com Atenção e Hiperatividade**. 2005. Dissertação (Mestrado) - Programa de Pósgraduação Stricto Sensu em Psicologia, Universidade São Francisco, Itatiba, São Paulo, 2005.

DAMASIO A. **Descartes' mistake**: emotion, reason and the human brain. São Paulo: Companhia das Letras, 1996.

DELIS, D. C.; KAPLAN, E. KRAMER, J. H. **Delis-Kaplan executive function system**. San Antonio, TX: The Psychological Corporation, 2001.

DELLA SALA, S. Pattern span: a tool for unwelding visuo-spatial memory. **Neuropsychologia**, v. 37, p. 1189-1199, 1999.

DELLA, SALA, S. et al. Assessing Dual-Task Performance Using a Paper-and-Pencil Test: Normative Data. **Archives of Clinical Neuropsychology**, v. 25, p. 410-419, 2010.

_____. Dual Task paradigm: a means to examine the central executive. **Annals of the New York Academy of Sciences**, v. 769, p. 161-171, 1995.

_____. Testing central executive functioning with a pencil-and-paper test. In: RABBITT, P. M. (Ed.). **Methodology of frontal and executive function.** Psychology Press: Hove, 1997.

DIAS, N. M. **Avaliação neuropsicológica das funções executivas**: Tendências desenvolvimentais e evidências de validade de instrumentos. 2009. Dissertação (Mestrado) - Programa de Pós-graduação em Distúrbios do Desenvolvimento. Mackenzie Presbyterian University, São Paulo, 2009.

DIAS, N. M.; MENEZES, A.; SEABRA, A. G. Alterações das Funções Executivas em Crianças e Adolescentes. **Estudos Interdisciplinares em Psicologia**, Londrina, v. 1, n. 1, p. 80-95, jun. 2010.

DIAS, N. M. et al. Age Differences in Executive Functions. **Spanish Journal of Psychology**, Universidad Complutense de Madrid and Colegio Oficial de Psicólogos de Madrid, v. 16, n. 9, p. 1-14, 2013.

DUNCAN, J. An adaptive coding model of neural function in prefrontal cortex. **Nature Reviews Neuroscience,** v. 2, p. 820-829, Nov. 2001.

_____ et al. A neural basis for general intelligence. **Science**, n. 289, p. 457-460, 2000.

EMSLIE, H. et al. **Behavioural Assessment of the Dysexecutive Syndrome for Children (BADS-C)**. London, UK: Harcourt Assessment/The Psychological Corporation, 2003.

FERNANDES, M. S.; MOSCOVITCH, M. Divided attention and memory: Evidence of substantial interference effects at retrieval and encoding. **Journal of Experimental Psychology**: General, v. 129, p. 155-176, 2000.

FISK, J. E.; SHARP, C. A. Age-related impairment in executive functioning: updating, inhibition, shifting, and access. **J Clin Exp Neuropsychol**, v. 26, p. 874-890, 2004.

FOURNIER-VICENTE, S.; LARIGAUDERIE, P.; GAONAC'H, D. More dissociations and interactions within central executive functioning: a comprehensive latent-variable analysis. **Acta Psychologica**, v. 129, n. 1, p. 32-48, 2008.

FRAY, P. J.; ROBBINS, T. W. CANTAB battery: proposed utility in neurotoxicology. **Neurotoxicol Teratol**. v. 18, n. 4, p. 499-50, Jul./Aug. 1996.

FRIEDMAN, N. P. et al. Individual differences in executive functions are almost entirely genetic in origin. **Journal of Experimental Psychology General**, v. 137, n. 2, p. 201-225, 2008.

_____. Greater Attention Problems During Childhood Predict Poorer Executive Functioning in Late Adolescence. **Association for Psychological**, v. 18, n. 10, 2007.

_____. Not All Executive Functions Are Related to Intelligence. **Psychological science**. v. 17, n. 2, 2006.

GATHERCOLE, S. E. et al. Working memory skills and educational attainment: Evidence from National Curriculum assessments at 7 and 14 years of age. **Applied Cognitive Psychology**, v. 18, n. 1, p. 1-16, 2004.

_____. Working memory skills and educational attainment: Evidence from National Curriculum assessments at 7 and 14 years of age. **Applied Cognitive Psychology**, v. 18, n. 1, p. 1-16, 2004.

HAMDAN, A. C.; PEREIRA, A. P. A. Avaliação neuropsicológica das funções executivas: Methodological considerations. **Psicol. Reflexo. Crit.**, Porto Alegre, v. 22, n. 3, 2009.

HAMDAN, A. C.; SOUZA, J. A.; BUENO, O. F. A. Performance of university students on random number generation at different rates to evaluate executive functions. **Arq Neuropsiquiatr**, v. 62, n. 1, p. 58-60, 2004.

HO, A. K. et al. Random number generation in patients with symptomatic and presymptomatic Huntington's disease. **Cog Behav Neurol**, v. 17, p. 208-212, 2004.

HO, S. C. et al. A 3-year follow-up study of social, lifestyle and health predictors of cognitive impairment in a Chinese older cohort. **International Journal of Epidemiology**, v. 30, n. 6, p. 1389-9136, 2001.

HUGHES, C.; RUSSEL, J.; ROBBINS, T. W. Evidence for executive dysfunction in autism. **Neuropsychologia**, v. 32, p. 477-492, 1994.

JAHANSHAHI, M. et al. The role of the dorsolateral prefrontal cortex in random number generation: a study with positron emission tomography. **Neuroimage**, v. 12, p. 713-725, 2000.

JURADO, M. B.; ROSSELLI, M. The elusive nature of executive functions: A review of our current understanding. **Neuropsychological Review**, v. 17, p. 213-233, 2007.

LEHTO, J. E.; JUUJÃRVI, P.; KOOISTRA, L.; PULKKINEN, L. Dimensions of executive functions: evidence from children. **British Journal of Developmental Psychology, v.** 21, n. 1, p. 59-80, Dec. 2010.

LEZAK, M. D.; HOWIESON, D. B.; LORING, D. W. **Neuropsychological assessment.** 4th ed. New York: Oxford University Press; 2004

LIMA, Ricardo Franco de; SALGADO, Cíntia Alves; CIASCA, Sylvia Maria. Desempenho neuropsicológico e fonoaudiológico de crianças com dislexia do desenvolvimento. **Rev. psicopedag.**, São Paulo , v. 25, n. 78, 2008.

LORING D. (Ed.). **INS Dictionary of neuropsychology**. New York: Oxford University Press, 1999.

LURIA, A. R. **Fundamentos de Neuropsicologia**. São Paulo: Scientific and Technical Books, 1981.

MIRANDA, M. C.; MUSZKAT, M. Neuropsychologia do Desenvolvimento. In: ANDRADE, V. M.; SANTOS, F. H.; BUENO, O. F. A. (Orgs). **Neuropsicologia Hoje** (211-224). São Paulo: Editora Artes Médicas Ltda, 2004.

MIYAKE, A. et al. The unity and diversity of executive functions and their contributions to complex frontal lobe tasks: a latent variable analysis. **Cog Psychol**, v. 41, p. 49-100, 2000.

_____. Inner speech as a retrieval aid for task goals: the effects of cue type and articulatory suppression in the random task cuing paradigm. **Acta Psychologica**, v. 115, n. 2-3, p. 123-142, 2004.

MIYAKE, A. et al. Inner speech as a retrieval aid for task goals: the effects of cue type and articulatory suppression in the random task cuing paradigm. **Acta Psychologica**, v. 115, n. 2-3, p. 123-142, 2004.

_____ FRIEDMAN, N. P. The nature and organization of individual differences in executive functions: four general conclusions. **Current Directions in Psychological Science**, v. 21, p. 8-14, 2012.

MORRIS, N.; JONES, D. M. Memory updating in working memory: The role of the central executive. **British Journal of Psychology**, v. 81, p. 111-12, 1990.

NATALE, L. L.; TEODORO, M. L. M.; HAASE, V. G. Avaliação neuropsicológica das funções executivas em crianças. In: ORTIZ et al. **Avaliação neuropsicológica panorama interdisciplinar dos estudos na normatização e validação de instrumentos no Brasil**. São Paulo: Editora Vetor, 2008.

NAVEH-BENJAMIN, M.; CRAIK, F. I. M.; GUEZ, J.; KREUGER, S. Divided attention in younger and older adults: Effects of strategy and relatedness on memory performance and secondary costs. **Journal of Experimental Psychology**: Learning, Memory, and Cognition, v. 31, p. 520-537, 2005.

NORMAN, D. A.; SHALLICE, T. Attention to action: Willed and automatic control of behavior. In: Davidson, R. J.; Schwartz, G. E.; Shapiro, D. (Eds.). **Consciousness and self regulation**. New York: Plenum Press, 1986. pp. 1-18.

NOVAK, G.; SOLANTO, M.; OSTROSKY-SOLÍS, F.; ARDILA, A.; ROSSELLI, M. NEUROPSI: a brief neuropsychological test battery in Spanish with norms by age and educational level. **Journal of International Neuropsychology Society**, v. 5, n. 5, p. 413-433, 1999.

OWEN, A. M. Cognitive planning in humans: neuropsychological, neuroanatomical and neuropharmacological perspectives. **Prog Neurobiol**, v. 53, p. 431-50, 1997.

PACKWOOD, S.; HODGETTS, H. M.; TREMBLAY, S. A multiperspective approach to the conceptualization of executive Functions. **Journal Of Clinical And Experimental Neuropsychology**, v. 33, n. 4, p. 456-47, 2011.

PAUL, W. B. et al. The case for the development and use of "ecologically valid" measures of executive function in experimental and clinical neuropsychology.
Journal of the International Neuropsychological Society, v. 12, p. 194-209, 2006.

PIAGET, J. **The construction of reality in the child**. New York: Basic Books, 1954.

PINHEIRO, A. M. V. **Contagem de frequência de ocorrência de palavras expostas a crianças na faixa de pré-escola e séries iniciais.** São Paulo: Brazilian Dyslexia Association, 1996.

PINHEIRO, A. M. V. **Contagem de frequência de ocorrência de palavras expostas a crianças na faixa de pré-escola e séries iniciais**. São Paulo (SP): Brazilian Dyslexia Association, 1996.

POMPÉIA, S.; MIRANDA, M. C.; BUENO, O. F. A. A set of 400 pictures standardised for Portuguese: norms for name agreement, familiarity and visual complexity for children and adults. **Arquivos de Neuropsiquiatria**, v. 59, n. 2-B, p. 330-337, 2001.

RABIN, L. A.; BARR, W. B.; BURTON, L. A. Assessment practices of clinical neuropsychologists in the United States and Canada: a survey of INS, NAN, and APA

Division 40 members. **Archives of Clinical Neuropsychology**, n. 20, p. 33-65, 2005.

REY, A. **Teste de cópia e de reprodução de memória de figuras geométricas complexas**: Manual. São Paulo: Casa do Psicólogo, 1999.

ROMINE, C. B.; REYNOLDS, C. R. A model of the development of frontal lobe functioning: Fin- dings from a meta-analysis. **Applied Neuropsychology**, v. 12, n. 4, p. 190-201, 2005.

SHALLICE, T. Specific impairments of planning. **Philos Trans R Soc Lond B Biol Sci**, v. 298, p. 199-209, 1982.

SIMÕES, C. et al. **Psychological assessment**: Instruments validated for the Portuguese population. Coimbra: Quarteto, 2013. (v. III, pp. 281-304).

SMITH, E. E.; JONIDES, J. Neuroimaging analyses of human working memory. **Proc Natl Acad Sci,** USA, v. 95, p. 12061-12068, 1998.

_____. Storage and executive processes in the frontal lobes. **Science**, v. 283, p. 1657-1661, 1999.

SPREEN, O.; STRAUSS, E. **A compendium of neuropsychological tests. Administration, norms, and commentary.** New York: Oxford University Press, 1998.

ST CLAIR-THOMPSON, H. L.; GATHERCOLE, S. E. Executive functions and achievements in school: Shifting, updating, inhibition, and working memory. **The quarterly journal of experimental psychology**, v. 59, n. 4, p. 745-759, 2006.

STROOP, J. R. Studies of interference in serial verbal reactions. **J Exp Psychol**, v. 18, p. 643-662, 1935.

STUSS, D. T.; BENSON, D. F. **Thefrontallobes**. NewYork, NY: Raven Press, 1986.

TAMNES, C. K. et al. Neuroanatomical correlates of executive functions in children and adolescents: a magnetic resonance imaging (MRI) study of cortical thickness. **Neuropsychologia**, v. 48, n. 9, p. 2496-508, Jul. 2010.

TOWSE, H. N.; MCLACHLAN, A. An exploration of random generation among children. **British Journal of Developmental Psychology**, v. 17, p. 363-380, 1999. TOWSE, J. N.; NEIL, D. Analyzing human random generation behavior: a review of methods used and a computer program for describing performance. **Behav Res Meth Instrum Comput**, v. 30, p. 583-591, 1998.

UEHARA, E.; CHARCHAT-FICHMAN, H.; LANDEIRA-FERNANDEZ, J.; Executive functions: An integrative portrait of the main models and theories of this concept. **Revista Neuropsicologia Latinoamericana,** v. 5, n. 3, p. 25-37, 2013.

VAN DER VEN, S. H. et al. The development of executive functions and early mathematics: a dynamic relationship. **The British Journal of Educational Psychology**,v. 82, pt. 1, p. 100-119, 2012.

_____. The development of executive functions and early mathematics: A dynamic relationship. **British Journal of Educational Psychology**, v. 82, p. 100-119, 2012.

WECHSLER, D. **Escala de Inteligência Wechsler para Crianças.** 3. ed. São Paulo: Casa do Psicólogo, 2002. 322 p.

WELSH, M. C.; FRIEDMAN, S. L.; SPIEKER, S. J. Executive functions in developing children: Current conceptualisations and questions for the future. In: MCCARTHY, K.; PHILLIPS, D. (Eds.). **Blackwell Handbook of Early Child Development**. Oxford: Blackwell, 2006.

WELSH, M. C.; PENNINGTON, B. F. Assessing frontal lobe functioning in children: views from developmental psychology. **Developmental Neuropsychology**, n. 4, p. 199-230, 1988.

WILSON, B. A. et al. **Behavioural Assessment of the Dysexecutive Syndrome**. Bury St Edmunds: Thames Valley Test Company, 1996.

_____ . **Behavioural Assessment of the Dysexecutive Syndrome.** Bury St Edmunds: Thames Valley Test Company, 1996.

YNTEMA, D. B.; SCHULMAN, G. M. Response selection in keeping track of several things at once. **Acta Psychologica**, v. 27, p. 325-332, 1967.

ZELAZO, P. D. et al. Early development of executive function: a problem-solving framework. **Review of General Psychology**, v. 1, n. 2, p. 198- 226, 1997.ZELAZO, P. D.; CRAIK, F. I. M.; BOOTH, L. Executive function across the life span. **Acta Psychologica**, v. 115, p. 167-184, 2004.

_____ ; QU, L.; MULLER, U. Hot and cool aspects of executive function: Relations in early development. In: SCHNEIDER, W.; SCHUMMANN-HENGSTELER, R.; SODIAN, B. (Eds.). **Young children's cognitive development**: Interrelationships among executive functioning, working memory, verbal ability, and theory of mind. Mahwah, NJ: Lawrence Eribaum Associates Publishers, 2004. p. 71- 93.

ANNEXES

ANNEX A - Questionnaire on clinical history and physical state

Parents,

In continuation of the research entitled *The unity and diversity of Executive Functions*, which you authorized your child to participate in, I request that the data below be filled out.

Any questions, call the Responsible Researcher:

Who filled in: ________________

Degree of kinship with the child: ______________________

Home Phone: _______________Mobile: ___________ Work: ______________

Full address:____

Full name of child: ____

Father's name: ____

Mother's name: ____

Child's date of birth: _______________ Age: __________

Gender: male ☐ female ☐

Weight: _____ Height: ________ BMI (weight/height): [2] _______

Is the child left-handed? yes ☐ no ☐

Is Portuguese your first language? yes ☐ no ☐ Do you speak another language? yes ☐

no ☐

Which one? ______________

Schooling: What grade are you in? Have you repeated any grade? yes ☐ no ☐

Which one? ______________

At what age did the child start school? did he or she attend pre-school? yes ☐ no ☐

Does the child have or had any problems learning to read and write? yes ☐ no ☐

Which one? ______________

How do you rate the child's school performance? Poor ☐ Fair ☐ Good ☐ Very good ☐

Good ☐ Great ☐

What is the child's greatest difficulty? Reading ☐ Writing ☐ Mathematics ☐

Other ☐ _______________________

Do you consider the child to be in good health? yes ☐ no ☐

Does the child live or has lived with someone who smokes indoors? yes ☐ no ☐

Does the child live or has lived with someone who ingests alcohol within the home? yes ☐

no ☐ How often?

Does the child ingest caffeine (coffee, tea, chocolate, soft drinks) daily? yes ☐ no ☐ How much per day?

Does the child sleep well? yes ☐ no ☐ When the child is not sleep deprived, how difficult is it to wake up at 7am? Answer on a scale from 0 (wakes up easily) to 10 (very difficult):

____

Does the child practise any physical activity? yes ☐ no ☐ Does it feel good when it exercises? yes ☐ no ☐

Is the child undergoing any treatment (medical, psychologist, speech therapist)? yes ☐ no ☐ Which?__________________

Is the child taking any medication? yes ☐ no ☐ Which?

Has the child taken any medication for a long period of time? yes ☐ no ☐

Which one? ________________

For how long did you take the medication? ________________________ For which why did you take the medication? ______________________________ Did you stop the

medication for how long? __________________________________

Has the child ever been unwell after taking a medicine? yes ☐ no ☐ If yes, which medicine and what symptom?

Does the child have vision: normal ☐ corrected ☐ If corrected, what is the problem, degree and correction? _______________

Does the child have or have had any hearing problems? yes ☐ no ☐ Does he/she wear a hearing aid? yes ☐ no ☐ Has he/she ever had frequent ear pain (otitis)? yes ▫ no ☐

Does the child have or had any difficulty producing or understanding speech? yes ☐ no ☐

Does the child sleep well? yes ☐ no ☐ If not, why not?

Does the child snore very loudly? yes ☐ no ☐ Does he/she have sleep apnea? yes ☐ no ☐ don't know ☐ Has anyone ever observed that the child stops breathing while sleeping? yes ☐ no ☐ Obs. about sleep/apnea _________________

Has the child been feeling very depressed or anxious lately? yes ☐ no ☐

Overall, do you consider the child more depressed or anxious than other children? yes ☐ no ☐

If yes, do you think this may reflect some pathology? yes ☐ no ☐

The child has had or has:

Diabetes? yes ☐ no ☐

Insulin resistance? yes ☐ no ☐

Hypoglycaemia? yes ☐ no ☐

Tumours? yes ☐ no ☐

Reading, learning, attention or hyperactivity difficulties? yes ☐ no ☐

Glaucoma? yes ☐ no ☐

Hypertension? yes ☐ no ☐

Liver problems (including hepatitis)? yes ☐ no ☐

Respiratory problems (including tuberculosis)? yes ☐ no ☐

Kidney problems? yes ☐ no ☐

Heart problems? yes ☐ no ☐

Hormonal (including thyroid) problems? yes ☐ no ☐

Neurological problems? yes ☐ no ☐

Have you ever fainted? yes ☐ no ☐

Have you ever had seizures, including epilepsy? yes ☐ no ☐ If yes, since what age?

Psychiatric problems, including depression? yes ☐ no ☐ Which ones?

Any allergic reaction? yes ☐ no ☐ To what?

Have you ever had any surgery? yes ☐ no ☐ Which?

Have you ever been hospitalised? yes ☐ no ☐ For how long?

Have you ever had a serious accident? yes ☐ no ☐ Describe:

Any other health problems? ____

Socio-economic information

Who is the head of the family in the child's household? ☐ Father ☐ Mother ☐ Other

What is the occupation of the head of the household? __Occupation: _________

What is the mother's (or guardian's) education?

☐ Illiterate/1ª to 4ª incomplete series - Last series attended:

☐ 1st to 4th grade (primary or primary schools I)

☐ 5ª to 8ª incomplete series - Last series attended:

☐ Complete 5th to 8th grade (junior high school or high school II)

☐ 1st to 3rd years incomplete - Last year attended:

☐ Complete 1st to 3rd year (high school, scientific or secondary school) / technical course - which? _______

☐ Incomplete higher education - How many years attended:

☐ Higher education completed

What is the education of the father (or guardian)

☐ Illiterate/incomplete to 4th grade - Last grade attended:

☐ 1ª to 4th grade complete (primary or primary schools I)

☐ Incomplete 5th to 8th grade - Last grade attended:

☐ Complete 5th to 8th grade (junior high or high school II)

☐ 1st to 3rd years incomplete - Last year attended:

☐ Complete 1st to 3rd year (high school, scientific or secondary school)/technical course - Which?

☐ Incomplete higher education - How many years attended:

☐ Higher education completed

Which and how many of these items does your family possess?

Colour TV: ___________ Videocassette/DVD: ___________Radio: ___________
Bathroom: _______

Car: _____Employee: ____________________Washing machine: ___________
Refrigerator: _____

Freezer (separate or 2nd door of the refrigerator): ____

Information referring to the day of the experiment:

Did you bring child's glasses/lens? yes ☐ no ☐ not necessary ☐

The child slept for how many hours yesterday? ____________________________

Did the child use any medication yesterday or today? yes ☐ no ☐ how much/what?

Did the child eat/drink anything before coming to the laboratory? yes ☐ no ☐

What? ____

When was the last time the child consumed caffeine (tea, coffee, chocolate, *cappuccino*, headache medicine)? ____

Name: _______________________________________ Date: ______________

I answered this form which consists of six pages.

Signature: ____

ANNEX B - Free and Informed Consent Form (FICF)

Dear Parents or Guardians

1 - "THE UNITY AND DIVERSITY OF EXECUTIVE FUNCTIONS: THEIR DISTINCTION AND INTER-RELATION IN THE ASSESSMENT OF CHILDREN WITH ATTENTION DEFICIT AND HYPERACTIVITY DISTRICT" is a research project of the Federal University of São Paulo, Guarulhos Campus. The researcher in charge is Andreia Cristina Correia, a Master's student in the Postgraduate Program in Education and Health in Childhood and Adolescence (Unifesp- Guarulhos), under the guidance of Prof. Dr. Sueli

Rizzutti.

2 - This information is being provided for voluntary participation in this study, where the executive functions in children with Attention Deficit Hyperactivity Disorder (ADHD) will be investigated through a neuropsychological evaluation battery.

3 - Two evaluation sessions will be held, one with You (father, mother or responsible for the child), of approximately one hour's duration, and another session with your child of approximately two hours' duration, in which questionnaires, scales and psychological tests measuring the child's cognitive and behavioural performance, quality of life and general health will be administered. With You, scales will be administered to provide information about behaviour

of your child and his/her health condition/history. You will also be sent scales to be completed by your child's teachers, assessing their classroom behaviour and academic performance.

4 - The tests and assessment procedures will be applied in a private room at the Núcleo de Atendimento Neuropsicológico Infantil Interdisciplinar (Nani) premises, being carried out by professionals specially trained for the purposes of the research.

5 - We clarify that the possible risks of the research are related to the initial difficulties of exposing their problems regarding their child's behavior and the anxiety inherent to every evaluation process. However, we guarantee that the discomforts will be mediated by the psychologists that make up the research team.

6 - Possible benefits for You are the neuropsychological assessment itself, which will provide information on the child's cognitive and behavioural performance, the preserved or compromised cognitive functions that impact on your child's academic performance, the identification of the presence or absence of behavioural problems, with suggestions for possible referrals for appropriate assessment and treatment.

7 - At any stage of the study, you will have access to professionals responsible for the research to clarify any doubts. The researchers can be reached at the Núcleo de Atendimento Neuropsicológico Infantil Interdisciplinar, Centro Paulista de Neuropsicologia, Rua Embaú, 54, Vila Clementino, São Paulo, or by telephone 11 5549-6899. If you have any considerations or questions about the ethics of the research, please contact the Research Ethics Committee (CEP), Rua Botucatu, 572 - 1o andar - Conjunto 14, phones 5571-1062, 5539-7162 (fax), *email*: cepunifesp@unifesp.br.

8 - The freedom to withdraw consent at any time is guaranteed and to stop participating in the study without any prejudice to the continuity of their follow-up at the Núcleo de Atendimento Neuropsicológico Infantil Interdisciplinar (Nani).

9 - The information obtained will be analysed together with that from other volunteers, with the identification of any participant not being disclosed.

10 - You will have the right to be kept up to date on the partial results of the research or information about your child that is known to the researchers.

11 - Should You agree to participate in this research, there are no personal expenses for the participant at any stage of the study, including examinations and consultations. There is also no financial compensation related to your participation. If there is any additional expense, it will be absorbed by the research budget.

12 - In case of personal injury, directly caused by the procedures or treatments proposed in this study (causal link proven), the participant is entitled to medical treatment at the Institution, as well as to compensation.

13 - The data and material collected will be used solely for this research. We clarify that the results of this research are for scientific and social purposes, in order to bring improvements in the quality of life of participants and health care practices and will be disclosed in events such as congresses, symposia, seminars and publication of results in journals, scientific journals, books, articles, among others.

14 - This consent form is printed in two copies, with one copy will remain with the responsible researcher and the other will be provided to you.

All pages must be initialled by the signatory.

I believe I have been sufficiently informed about the research, through the term that I have read or that was read to me, describing the study. I agree with the above mentioned items. I agree with the procedures to be performed, their discomforts as described above, as well as the risks and the guarantees of confidentiality. I voluntarily agree to participate and/or accompany the one I am legally responsible for and that I may withdraw my consent at any time before or during the process, without penalty or prejudice or loss of any benefit that I or the one I care for may have acquired through the research.

I authorise the provision of data on the child's assessment, if requested by the school:

( ) Yes ( ) No

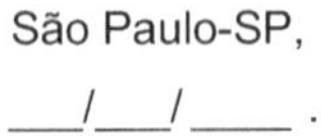

São Paulo-SP,

___/___/____ .

NAME OF THE PERSON RESPONSIBLE FOR THE CHILD

Degree of kinship ____________________

Signature of the person responsible for the child

Researcher in charge

If you have any questions about the ethical aspects of this study, you may consult:

RESPONSIBLE RESEARCHER: ANDREIA CRISTINA CORREIA

Rua Embaú, 54, Vila Clementino - São Paulo - Brasil - CEP 04039-060

(55-11) 5549-8476/6899 | FAX (55-11) 5576-5092

andreia.correia@unifesp.br

www.unifesp.br/dpsicobiolo/gruponani www.cpsnp.com.br

ANNEX C - Term of Consent

You are being invited as a volunteer to participate in a pilot study to test a battery of cognitive functions that is being developed by a team of researchers from the Federal University of São Paulo, Department of Psychobiology and Department of Education and Health in Childhood and Adolescence.

This application is part of the scientific work which intends to analyse the suitability of tasks to assess executive functions (i.e. which involve behavioural inhibition skills, organisation, planning, flexibility, self-monitoring, among others) and to verify their sensitivity in detecting cognitive deficits in childhood.

We are doing this pilot project with children your age (6 years old), who live here in São Paulo. To participate in this study, your parents or guardians must authorise and sign a document stating that they accept your participation. You can choose whether or not to participate. We have discussed this research with your parents or guardians and they know that we are also asking for their agreement. If you are going to take part in the research, your parents or guardians will also have to agree. But if you do not want to take part in the research, you are not obliged to, even if your parents agree. There will be no cost to you, and you will not be paid any money for taking part. You will be able to ask questions about anything you want, and you will be free to take part or not. Your parents or guardians will be able to stop, i.e. withdraw your participation in the research whenever they wish. Your participation is voluntary and if you do not wish to take part there will be no problems. We will not tell other people that you are in this research and we will not share information about you with anyone who is not working on the research. After the research is completed, the results will be reported to you and your parents and may be published in scientific papers without revealing your name. You will not be named in any such publication or release to the public. This study presents minimal risk, that is, the same risk that exists in your day to day activities, such as talking, doing homework, reading, etc. Your name or material indicating your participation will not be released without permission from the person responsible for you. All research material concerning you will be stored with the responsible researcher for a period of 5 years, after which time it will be destroyed. This consent form is printed in duplicate, one copy will remain with the responsible researcher and the other will be given to you.

I, ____, have been informed of the objectives of the present study in a clear and detailed manner and clarified my doubts. I know that at any time I may request new information, and my guardian may modify the decision to participate if he/she so wishes. Having the consent of my guardian already signed, I declare that I agree to participate in this study. I received a copy of this consent form and was given the opportunity to read it or it was read to me and clarify my doubts.

São Paulo, ___ from __________________ 20__________

Signature of minor

Signature of the researcher

If you have any questions about the ethical aspects of this study, you may consult:

RESPONSIBLE RESEARCHER: *ANDREIA CRISTINA CORREIA*

Rua Embaú, 54, Vila Clementino - São Paulo - Brasil - CEP 04039-060

(55-11) 5549-8476/6899 | FAX (55-11) 5576-5092 andreia_cristina10@yahoo.com.br

ANNEX D - School Statement

PILOT STUDY - NEUROPSYCHOLOGICAL ASSESSMENT OF EXECUTIVE FUNCTIONS

São Paulo, 2013

Dear Sir or Madam

My name is Andreia Cristina Correia, I am a psychologist and I would like to do a pilot study to test a battery of cognitive functions that is being developed by a team of researchers from the Federal University of São Paulo, Department of Psychobiology. This battery will be applied to 10 children aged 6 years old.

This application is part of the scientific work that intends to analyse the suitability of tasks to assess executive functions (i.e. those that involve behavioural inhibition, organisation, planning, flexibility, self-monitoring skills, among others) and to verify their sensitivity in detecting cognitive deficits in childhood.

Children will be selected from I° to 5° year, compatible with the school year. These children will take a letter to their parents inviting one of them to attend the school to be interviewed and give consent to apply the test in the child. After the return of these documents, the child will be evaluated at the school. I need a quiet room for the application of the test. There will be two sessions lasting approximately 60 minutes each.

This study was approved by the Research Ethics Committee of the Federal University of São Paulo (UNI FESP).

Sincerely,

Master's student - Andreia Cristina Correia

Responsible Researchers: Prof. Dra Monica C Miranda and Prof. Dra Sabine Pompéia Universidade Federal de São Paulo

Autorizado por
Nome:
Cargo/Função____________________
Assinatura

ANNEX E - Approval of the Committee

UNIVERSIDADE FEDERAL DE
SÃO PAULO - UNIFESP/
HOSPITAL SÃO PAULO

PARECER CONSUBSTANCIADO DO CEP

DADOS DO PROJETO DE PESQUISA

Título da Pesquisa: A UNIDADE E A DIVERSIDADE DAS FUNÇÕES EXECUTIVAS: SUA DISTINÇÃO E INTER-RELAÇÃO NA AVALIAÇÃO DAS CRIANÇAS COM TRANSTORNO DE DÉFICIT DE ATENÇÃO E HIPERATIVIDADE

Pesquisador: ANDREIA CRISTINA CORREIA DA SILVA

Área Temática:

Versão: 2

CAAE: 11984912.1.0000.5505

Instituição Proponente: Universidade Federal de São Paulo - UNIFESP/EPM

DADOS DO PARECER

Número do Parecer: 186.770

Data da Relatoria: 18/01/2013

Apresentação do Projeto:

Conforme parecer CEP 184.764 de 11/1/2013

Objetivo da Pesquisa:

Conforme parecer CEP 184.764 de 11/1/2013

Avaliação dos Riscos e Benefícios:

Conforme parecer CEP 184.764 de 11/1/2013

Comentários e Considerações sobre a Pesquisa:

Conforme parecer CEP 184.764 de 11/1/2013

Considerações sobre os Termos de apresentação obrigatória:

Foram apresentadas as respostas às pendencas apontadas. Foi incluido novo TCLE e novo termo de assentimento.

Recomendações:

não há.

Conclusões ou Pendências e Lista de Inadequações:

sem inadequações. Projeto pode ser liberado.

Situação do Parecer:

Aprovado

Endereço: Rua Botucatu, 572 1º Andar Conj. 14
Bairro: VILA CLEMENTINO **CEP:** 04.023-061
UF: SP **Município:** SAO PAULO
Telefone: (11)5539-7162 **Fax:** (11)5571-1062 **E-mail:** cepunifesp@unifesp.br

UNIVERSIDADE FEDERAL DE
SÃO PAULO - UNIFESP/
HOSPITAL SÃO PAULO

Necessita Apreciação da CONEP:

Não

Considerações Finais a critério do CEP:

As pendências foram atendidas, o colegiado acatou o parecer do relator. Projeto aprovado.

SAO PAULO, 18 de Janeiro de 2013

____

Assinador por:
José Osmar Medina Pestana
(Coordenador)

Endereço: Rua Botucatu, 572 1º Andar Conj. 14
Bairro: VILA CLEMENTINO **CEP:** 04.023-061
UF: SP **Município:** SAO PAULO
Telefone: (11)5539-7162 **Fax:** (11)5571-1062 **E-mail:** cepunifesp@unifesp.br

Printed by Books on Demand GmbH, Norderstedt / Germany